# Disorders of Human Communication 2

Edited by G.E. Arnold, F. Winckel, B.D. Wyke

# Clinical Aspects
of Dysphasia

Springer-Verlag Wien New York

Martin L. Albert, M. D.

Professor of Neurology and Clinical Director, Aphasia Research Center, Boston University Medical School, and Chief, Clinical Neurology Section, Boston Veterans Administration Medical Center, Boston, Mass., U.S.A.

Harold Goodglass, Ph. D.

Professor of Neurology (Neuropsychology), Boston University Medical School, and Director, Aphasia Research Center, Boston University Medical School, and Director, Psychology Research, Boston Veterans Administration Medical Center, Boston, Mass., U.S.A.

Nancy A. Helm, D. Sc.

Assistant Professor of Neurology (Speech Pathology) and Director of Audiology/Speech Pathology Program, Neurology Service, Boston Veterans Administration Medical Center, and Boston University Medical School, Boston, Mass., U.S.A.

Alan B. Rubens, M. D.

Associate Professor of Neurology, University of Minnesota Medical School, and Director, Neurobehavior Unit, Hennepin County Medical Center, and Assistant Chief of Neurology, Hennepin County Medical Center, Minneapolis, Minn., U.S.A.

Michael P. Alexander, M. D.

Assistant Professor of Neurology, Boston University Medical School, and Chief, Neurobehavior Unit, Boston Veterans Administration Medical Center, Boston, Mass., U.S.A.

With 12 Figures

Library of Congress Cataloging in Publication Data. Main entry under title: Clinical aspects of dysphasia. (Disorders of human communication ; 2.) Bibliography: p. Includes index. 1. Language disorders. I. Albert, Martin L. II. Series. [DNLM: 1. Aphasia. W1 D1762 v. 2 / WL 340.5 c641.] RC423.C55. 616.85'52. 81-4483

ISSN 0173-170X
ISBN-13:978-3-7091-8607-7        e-ISBN-13:978-3-7091-8605-3
DOI: 10.1007/978-3-7091-8605-3

# Editors' Foreword

This volume is one in a series of monographs being issued under the general title of "Disorders of Human Communication". Each monograph deals in detail with a particular aspect of vocal communication and its disorders, and is written by internationally distinguished experts. Therefore, the series will provide an authoritative source of up-to-date scientific and clinical information relating to the whole field of normal and abnormal speech communication, and as such will succeed the earlier monumental work "Handbuch der Stimm- und Sprachheilkunde" by R. Luchsinger and G. E. Arnold (last issued in 1970). This series will prove invaluable for clinicians, teachers and research workers in phoniatrics and logopaedics, phonetics and linguistics, speech pathology, otolaryngology, neurology and neurosurgery, psychology and psychiatry, paediatrics and audiology. Several of the monographs will also be useful to voice and singing teachers, and to their pupils.

**G. E. Arnold,** Jackson, Miss.
**F. Winckel,** Berlin
**B. D. Wyke,** London

# Preface

Neurologists, neuropsychologists, speech pathologists and other clinicians who care for dysphasic patients have often complained that available books on dysphasia tend to be parochially theoretical, and insufficiently directed towards clinical reality. These books provide the categories, labels, and theoretical speculations of one school or another; but dysphasic patients as often as not do not fit neatly into a specific theoretical category. Clinical patterns of dysphasic syndromes of most patients with dysphasia rarely conform fully to the pictures painted in the textbooks. This clinical reality is especially puzzling to those new to the field, who have assiduously learned the theories and labels, and who now expect dysphasic patients to comply. How often has the experienced clinician heard the student ask, "But what shall I call it? I have to put something down on the report."

What is needed then is not another book describing the syndromes of dysphasia but a book describing an "approach" to the dysphasic patient. In such a book the emphasis would be shifted from a concern for the name of a dysphasic syndrome to a concern for the patient. Key questions would be: what are the clinical manifestations of this patient's problems? what neurological disruptions underlie these problems? what preserved pockets of strength does this patient have which may be called on for rehabilitation? and what therapeutic approaches are available for such a patient?

In this book we have attempted to relate current trends in aphasiological research to the needs of clinicians for the benefit of their patients. Specifically, this book was designed around two issues: 1. What clinical approaches may be useful in dealing with dysphasic patients? and why? 2. Once a diagnosis of dysphasia has been made, what can be done? and why?

Although the work of others is summarized and discussed, the emphasis here is on our own clinical and research experience. Bedside testing techniques as well as more formal diagnostic approaches are provided. Syndromes of dysphasia are presented in a manner as to relate contemporary neuropsychological and neurolinguistic knowledge to what is actually seen at the bedside. Approaches to therapy are described which derive directly from the clinical symptomatology and which lead to specific functional goals.

A unique form of collaboration was developed for this book—unique in the sense that so many authors worked together to co-author a single monograph on dysphasia. This book was not developed as a collection of independent, unrelated segments. Although each co-author was primarily responsible for writing the initial draft of a particular portion, each was also responsible for reviewing all sections. We have attempted to avoid providing just another review of the literature. Our goal has been to write a comprehensive, integrated, palatable guide for the clinician—a guide which is not only informative on a theoretical level but also practical for every day use.

We are indebted to our many colleagues for constructive comments and criticisms in conversations and at meetings over the years. Our current opinions derive not only from our own clinical observations and research but also from interaction with and reaction to ideas and approaches of our colleagues. In particular, we wish to thank Professor Norman Geschwind who suggested that we undertake this project and who continually encouraged us in it.

Boston, Mass., and Minneapolis, Minn.,    **Martin L. Albert**
March 1981             **Harold Goodglass**
                    **Nancy A. Helm**
                    **Alan B. Rubens**
                    **Michael P. Alexander**

# Contents

# Part I
# Examination of the Dysphasic Patient

# General Clinical Considerations

Language is a means by which people communicate with each other using verbal symbols. Dysphasia may be defined as a disorder of language due to brain damage. This monograph deals with disorders of language, not disorders of speech. Speech refers to the mechanical process of articulation, which can be disturbed by weakness, slowness, or incoordination of the muscles of the glossopharyngeal apparatus. Such disturbances would be termed dysarthria, dysphonia, or mutism. The term dysphasia is applied to a neurological disorder resulting from damage to those regions of the cerebral hemispheres which form the anatomical basis for the human capacity for language.

No accurate estimate of the incidence or prevalence of dysphasia currently exists, to our knowledge. To obtain such an estimate would be a formidable task, since dysphasia may accompany such a wide variety of neurologic disorders, including head injuries, tumors, strokes, and infections. A rough estimate of incidence of dysphasia following stroke may be calculated, however. Brust *et al.* (1976) reported that 21 percent of 850 patients seen by them following acute onset of stroke were dysphasic. Kurtzke, in his 1969 study on the epidemiology of stroke, estimated the incidence of stroke in the United States to be 207 per 100,000 per year, or about 400,000 new strokes per year. Brust *et al.* then calculated that 21 percent of these 400,000 new stroke victims per year would be dysphasic—or 84,000 new cases per year from stroke alone.

The clinician examing a dysphasic patient has several specific goals in mind, including answers to the following questions: Which parts of the brain are damaged? What is the nature of the lesion (*e.g.* vascular, infectious, etc.)? What kind of dysphasia is present and what is its pathophysiologic basis? Which parts of the brain are spared and can these healthy regions of the brain be utilized to compensate for lost verbal abilities? The basic clinical aim, then, is a search for some neurobehavioral mechanism by which the dysphasic patient can communicate.

A formal language evaluation can provide detailed answers to these questions. Such an examination, however, may take from two to twelve hours, depending on the nature of the dysphasic deficit, and does not provide the busy clinician with a quick guide to the diagnosis from which an initial series of management steps may be undertaken. For this purpose a brief examination, as described below, for dysphasia can be used. This brief examination can be completed in fifteen minutes, can be carried out at the bedside with no need for special testing equipment beyond a pencil and paper, and can provide a general guide to initial diagnosis and treatment. The same examination, if followed systematically, can also be used on a daily basis to monitor the course and progression of the dysphasic syndrome.

## An Approach to Examining the Dysphasic Patient

Some basic items of medical history are necessary in the investigation of language disorders. *Handedness* of the patient should always be ascertained. Over 95 percent of righthanders and about 60 percent of lefthanders have language organized in the left hemisphere. For the remainder, either the right hemisphere is dominant, or language is organized bilaterally. The *native language* of the dysphasic patient should be determined; there is suggestive evidence that language may not be organized in the brain of a bilingual in the same manner as in that of a monolingual. Finally, the examiner should know the *level of education* of the patient, since linguistic performance depends on level of academic attainment.

In the clinical approach to the dysphasic patient, the examiner should use all available clues to diagnosis, whether they are linguistic or not. Evidence of neurological disease other than the language disorder can be helpful in determining the nature of the dysphasia. Presence of a significant hemiplegia places the lesion in motor pathways and suggests that serious impairment in spontaneous speech production will be present; and that the dysphasic syndrome will be of a non-fluent type. Presence of a significant hemisensory defect or homonymous hemianopia, in the absence of hemiplegia, suggests that the dysphasic syndrome will have been caused by a more posteriorly located lesion and that the language disorder is likely to be of a fluent type. Presence of all three—hemiplegia, hemisensory deficit, hemianopia—is most likely to be associated with a mixed or global dysphasia.

In evaluating the language disorder itself, the examiner should consider oral and written language separately. Useful bedside tests of oral language should include a sampling of spontaneous speech, repetition, naming, and comprehension. Tests of written language should sample reading and writing.

For purposes of gross clinical diagnosis and clinico-anatomic correlation, three major distinctions may be made. The first is between fluent and non-fluent speech output. Non-fluent speech is slow, laboriously produced, with abnormal speech rhythm and melody, poor articulation, shortened phrase length, and preferential use of substantive words (such as nouns and main verbs) rather than grammatical words (such as conjunctions and auxiliary verbs). Non-fluent speech, often called telegraphic or agrammatic, is frequently associated with anteriorly located lesions and is usually a feature of the anterior dysphasias. Fluent speech is produced at a normal or hypernormal rate, with normal speech rhythm and melody, good articulation, and normal or hypernormal phrase length. When fluent speech is part of a dysphasic syndrome, the lesion is usually located posteriorly in the cerebral hemisphere, and the syndrome is called a posterior dysphasia.

The second major clinical distinction is between presence or absence of repetition defect. When repetition defect is present the lesion is usually within the "zone of language" (Dejerine, 1914). The zone of language is in the bed territory of the middle cerebral artery, includes both banks of the sylvian fissure, and incorporates Broca's area anteriorly, Wernicke's area posteriorly, and the arcuate fasciculus between them. Dysphasic syndromes with repetition defect include Broca's, Wernicke's, and conduction dysphasia. The middle cerebral artery or one of its branches is often occluded in the production of a dysphasic syndrome with repetition defect. When repetition defect is absent the lesion has usually spared the zone of language and is located cortically or subcortically outside of, but near, the zone of language. The cortical lesions would be in the border zone between end arteries of the middle cerebral-anterior cerebral groups and the middle cerebral-posterior cerebral groups. Syndromes associated with such cortical lesions include transcortical motor, transcortical sensory dysphasia, isolation of the speech area, and anomic dysphasia. A likely cause would be thrombosis of the internal carotid artery. Subcortical lesions in the thalamus or basal ganglia may also produce dysphasic syndromes without repetition defect.

The third major clinical distinction is between disorders of oral language and disorders of written language. When the cerebral damage is vascular in etiology, and the language problems affect oral language only, the lesion is usually in the distribution of the middle cerebral or internal carotid artery. When the language problem affects written language only, the lesion is usually in the distribution of the posterior cerebral artery. When both spoken and written language are affected, more often than not the lesion, if there is only one, is in the carotid circulation.

# B Brief, Clinical (Bedside) Examination for Dysphasia

In a short examination for dysphasia six language skills should be tested: spontaneous speech, repetition, naming, comprehension of spoken language, reading, and writing. The following examination can be completed in 10—15 minutes at the bedside.

## 1. Spontaneous Speech

Spontaneous speech can be elicited by conversation with the patient; often this aspect of the examination will be completed during the taking of the medical history. Both the form and the content of language output should be evaluated. Form refers to features of fluency or non-fluency, as discussed above, e.g. effort to produce speech, rate of speaking (normal average is 100—125 words per minute), melody, phrase length (normal average is 3—5 words strung together between pauses).

Content refers to features of word choice, syntax, and presence or absence of paraphasias (described below). As to word choice, a patient with an anterior dysphasia is likely to use few highly meaningful, substantive words. A patient with a posterior dysphasia is more likely to be circumlocutory, using many words to talk around a subject without precision. In such patients there may be an excessive drive to continue speaking (press of speech, logorrhea). As to syntax, the anterior dysphasic will often be agrammatic, misusing or even completely avoiding the use of small grammatical or filler words. With respect to paraphasias, or verbal substitutions, these abnormalities of language may be seen in any variety of dysphasia. Phonemic paraphasias refer to the substitution of one correct phoneme for another (e.g. the patient says "pable" when "table" was intended). Semantic paraphasias refer to the substitution of a correct word for another (e.g. the patient says "orange" when "green" was intended). Neologistic paraphasias refer to the creation of new words which hadn't, to that moment, existed in the speaker's lexicon and which have no obvious meaning to the examiner (*e.g.* the patient says "I *mazmin* with him").

## 2. Repetition

The examiner utters the words to be repeated and asks the patient to "say what I say" or "repeat after me". Items to be tested include single words (*e.g.* object names, colors, numbers) and sentences of increasing length and syntactic complexity. Start with shorter items, especially items in which lip-reading can benefit the patient (*e.g.* "baby"), and progress to longer items, including words in which lip reading is of less obvious benefit (*e.g.* "ginger cake").

In testing repetition the examiner looks for the same production features as in testing spontaneous speech, *e.g.* features of fluency vs non-fluency, ease or difficulty of production of high content words vs grammatical words, presence and type of paraphasia, etc. Repetition may be defective, normal, or hypernormal (echolalia).

## 3. Naming or Word Finding

Impaired ability to name an object, color, body part, etc., or to find the desired word for production in spontaneous speech is present in every type of dysphasia. For purposes of a brief, clinical, bedside examination, it is sufficient, therefore, to document the existence of a naming or word-finding deficit. It is not necessary to detail the specific features of the naming defect; for this purpose, the formal language evaluation is more useful.

Word finding deficit may be detected in the examination of spontaneous speech. The patient may be circumlocutory, using many words to describe a thought which could be more simply explained in fewer words if these words were available to the patient.

Confrontation naming is tested by presenting a test stimulus — object, picture, color, etc. with the request to "tell me what this is". Special equipment is not necessary; any items in the room or on the person of examiner or patient are suitable. Patients with mild naming disorders may have the disorder brought out more easily by testing object parts, as well as the whole object (*e.g.* if the patient can name the target stimulus "watch", the examiner might ask for the names of "stem", "face", "hands", "crystal", etc.).

## 4. Comprehension of Spoken Language

In testing comprehension of spoken language the examiner must accept his or her dependence on the availability to the patient of a controllable output channel. If the patient were totally paralyzed, and had no way of controlling expression in any form, comprehension could not be tested. A first step in testing comprehension, then, should be the determination of a motor response which the patient can control in a consistent manner.

Experts often have difficulty in evaluating comprehension, for example, in the patient who has a severe dyspraxia.

Two approaches are generally successful: ask the patients to point to objects in the room, ask the patients questions which can be answered "yes" or "no". A series of questions of graded difficulty can then be presented. For example, a patient can be asked to "point to a light in this room" or to "point to a source of illumination." For the "yes-no" responses, questions might range from easy (*e.g.* "Is your name Dr. Smith?") to difficult (*e.g.* "Does the sun rise in the East?").

## 5. Reading

Reading aloud and reading comprehension should be tested separately, since these two language skills can be impaired independently in dysphasia. Reading aloud can be tested by presenting written material in script or block letter form. Tests of reading comprehension should take into account the same problem of output control discussed above in the section on comprehension of spoken language. Written names of common objects in the room can be shown to the patient who may demonstrate comprehension by pointing to the object. A series of questions of graded difficulty can be presented in written form to the patient, the examiner requesting a "yes" or "no" reply. A card with the words "yes" and "no" may be presented with the written question, and the patient may be requested to point to the answer, rather than say "yes" or "no" if the patient has difficulty with control of speech output. The examiner should pay particular attention to disorders of eye movements, hemianopia, or neglect, as possible non-linguistic causes of reading disorder.

## 6. Writing

In testing for dysgraphia, one should distinguish between disorders of a perceptual or mechanical nature (unsteadiness, poor alignment of letters in a word or of words in a sentence, poorly formed letters, slanted lines, neglect of one side of the line) and disorders of a linguistic nature (agrammatism, spelling errors, word substitutions). Writing disorders of a linguistic nature are common in dysphasic syndromes, and may be tested by asking the subject to write single letters and digits, words and multi-digit numbers, and sentences of increasing length and complexity. Writing to dictation may be tested independently of writing to command. In the latter test, the patient may be asked to demonstrate written confrontation naming ability (for comparison with spoken naming ability) or the patient may be given an open-ended task, such as "Write something about why you are in the hospital." Comparison with spontaneous speech may thus be made, and the same features as in spontaneous speech may be analyzed: fluency, content, etc.

# Clinical Guide to Classification of Dysphasic Syndromes   C

The preceding short examination for dysphasia can be used to provide a rough clinical guide to classification of dysphasic syndromes. Formal language evaluation should be carried out to refine the initial impression. We emphasize that many patients with dysphasia do not have signs which can be easily or neatly categorized, regardless of the technical skill or years of experience of the examiner; in such cases a thoughtful description of the observation is more helpful than an attempt to force the clinical findings to conform to a pre-conceived category.

## 1. Spontaneous Speech

Non-fluency is usually, but not always, associated with anterior lesions; fluency with posterior lesions. Anterior dysphasias include Broca's dysphasia and transcortical motor dysphasia. Posterior dysphasias include conduction, anomic, Wernicke's, and transcortical sensory dysphasia. The mixed varieties of dysphasia—isolation of the speech area and global dysphasia—are generally non-fluent. Posterior dysphasias often contain abundant phonemic and semantic paraphasias. The paraphasias of anterior dysphasias are more likely to be predominantly phonemic. Anterior dysphasia is characteristically agrammatic.

## 2. Repetition

Repetition defect is found in Broca's, conduction, Wernicke's, and global dysphasia. Repetition is relatively normal in anomic dysphasia and the dysphasias of subcortical lesions. In the transcortical dysphasias repetition may be normal or even hypernormal.

## 3. Naming

Disorders of naming are found in all varieties of dysphasia. When naming defects occur as the sole or major feature of dysphasia, we speak of anomic dysphasia.

## 4. Comprehension of Spoken Language

Obvious impairment of auditory comprehension is found in global dysphasia, the syndrome of isolation of the speech area, Wernicke's dysphasia, and transcortical sensory dysphasia.

## 5. Reading

Dyslexia may be found together with the anterior, mixed or posterior dysphasias; it may be associated with dysgraphia in the absence of disorders of spoken language; or it may appear as an isolated language defect.

## 6. Writing

Pure dysgraphia is extremely rare; some experts even doubt its existence as an isolated syndrome. Dysgraphia is commonly seen, however, in association with dyslexia or dysphasia.

# Formal Language Evaluation D

## 1. Five Aspects of Assessment in Dysphasia

The formal, clinical assessment of dysphasia encompasses at least five aspects. Depending on the circumstances, all may be given equal weight, one may be dominant to the exclusion of all the others, or any distribution of emphasis between these extremes may obtain. These aspects are the following: 1. dysphasia testing as an inventory of language input and output modalities, 2. linguistic aspects, 3. diagnostic aims, 4. the dysphasia examination as a case study, 5. quantitative aspects.

### Dysphasia Testing as an Inventory of Language Input and Output Modalities

By the very structure of the human communicative apparatus we are given the fact that language is emitted orally or graphically and perceived auditorily or in writing. Non-linguistic stimuli destined to elicit speech may impinge on the individual through the visual, auditory, tactile, or somesthetic senses. Moreover, the processing of a linguistic stimulus may lead to an output in the form of some motor behavior impinging on the environment.

It would seem logical to suppose then that a comprehensive dysphasia examination should be composed of a sample of the pairing of each possible input modality with each possible output. Indeed, a simple clinical examination on these principles was described by Chesher (1937). An elaboration of Chesher's approach might take the following form (Fig. 1).

This 4×4 matrix with 16 subtests could easily serve as the basis for an inventory of language functions. It has particular appeal to those who prefer to think of language as a collection of responses—each linked to a particular stimulus (*cf.* Wepman and Jones, 1961). For reasons to be developed later,

| Stimulus | Response | | | |
|---|---|---|---|---|
| | Point | Say | Write | Do |
| See Object | Visual matching | Naming | Written naming | Pantomine (Praxis) |
| Hear Words (Sentences) | Word discrim.<br><br>Sentence comprehension | Word repetition<br><br>Sentence repetition or answering questions | Writing from dictation | Follow commands |
| See Words (Sentences) | Word-object matching | Oral reading | Copy | Follow written commands |
| Feel Objects | Visual-tactile matching (Stereognosis) | Tactile naming | Tactile-written naming | |

Fig. 1. Stimulus response matrix

we do not believe that this extreme stimulus—response model of language is a valid one. However, the above matrix is a useful device for designing the contents of a dysphasia inventory because it suggests alternative input channels for testing a particular mode of output and vice versa.

Obviously, with such a matrix as its basis, an examination scaled in difficulty can be elaborated by varying the vocabulary and by introducing commands of increasing length and grammatical complexity.

Whatever the theoretical rationale underlying a particular examiner's orientation, he cannot escape incorporating most of the components of this matrix in his examination. The assessment of language function must deal with this aspect of dysphasia—its possible selectivity for particular modalities of input or output and—in some instances—for specific input-output combinations. However, such an examination matrix is based on purely logical, not clinical considerations. It could be devised by one who has never seen a dysphasic and knows nothing of its typology, of the variation in performance as a function of the communicative situation, nor of the many psycholinguistic variables which enter into the symptomatology and which should be specifically probed for. Moreover, some of the stimulus-response combinations (e.g. read-do) have proven to be unreliable and the corresponding capacities must be probed for by other means. In this section we will explore the many facets of dysphasia assessment which escape this simplified, though logical approach.

## Linguistic Aspects of Dysphasia Testing

If dysphasia could be summed up in terms of its effects on the various input and output channels of communication, it would be a simpler, but less fascinating problem for the clinician. The fact is that, within any modality, the breakdown of language may demand description in terms of the linguistic levels of organization which are most affected.

The chief levels of linguistic organization are the following:

a) *Phonological:* that level concerned (on the output side) with the generation of the individual sounds and syllables of the speakers language and (on the input side) with the discrimination and identification of these sounds, for the purpose of recognizing the words composed of them.

The phonological level may be analyzed further into two aspects: the *phonetic* and the *phonemic.* Phonetics is concerned with the relationship between the physical configuration of the articulatory apparatus at any moment during speech and the precise quality of the corresponding speech sound. Thus, articulatory disorders at the phonetic level are those which involve inadequate correspondence between the articulatory gesture and the target sound, resulting in effortful and inaccurate realizations of intended sounds. Actual observation of patients with this component of dysphasia suggests that their disorder is one of motor control and coordination. They may be unable to initiate speech movements at all or may fumble with totally inappropriate movements in their attempts to name or repeat. However, careful analyses of speech errors (Blumstein, 1973) reveals that a purely motor coordination interpretation is inadequate. Misarticulations are bound by linguistic rules, success varies as a function of the familiarity of the word and as a function of the availability of an auditory cue, which may be as little as the first sound of the intended word. The fact that a model for repetition usually facilitates articulation complicates the examination of this function, since it means that repetition of test words may not give a faithful reflection of the patient's actual clinical status. While repetition is certainly still one of the conditions under which articulation should be tested, we will be discussing other more representative approaches for characterizing this function quantitatively and qualitatively.

The second aspect of phonological function refers to the *phonemic level.* Phonemics refers to that problem in linguistics which identifies those attributes of speech sounds which are critical for determining meaning. Errors in production at the "phonemic" level, then, involve the mistargeting of the sound or syllable which is to be produced, rather than the inadequate rendition of a correctly chosen sound. While difficulties at the phonetic level, expressed in the form of awkward, labored articulation, tend to appear throughout a patient's speech, phonemic errors occur more unpredictably and sporadically. Since each instance represents a substitution of an unintended sound, they come under the heading of paraphasia—in this instance, "phonemic"—also called "literal" paraphasia. Literal paraphasias, if potentially present, are easy to elicit by repetition tasks of a mildly "tongue twisting" type—particularly those involving polysyllabic words with alternating place of consonant articulation, such as "basketball player" or "hippopotamus". Further details of examination procedures will be developed later in this section.

b) *Lexical:* The ability to articulate individual sounds or syllables may be totally independent of the ability to choose and produce an intended target word. The latter ability, dealing with word retrieval, is at the lexical level of function. The evaluation of the status of lexical function generally focuses on substantives, principal verbs, adjectives and adverbs, *i.e.* the "contentive" words—those which carry the informational load of the sentence. The grammatical words—articles, pronouns, prepositions and auxiliary verbs behave as though they are in another category. They are generally easily available to those patients (*i.e.* anomic dysphasics and Wernicke's dysphasics) whose primary problem is in word finding.

Naturally, the primary approach to word finding is by requiring naming to visual confrontation—with actual objects or pictures of objects, colors, actions, etc. However, this approach to testing does not discriminate between diagnostic groups of patients. The reason is that reduced access to the lexicon characterizes dysphasics of all types—it represents the severity, but not the type of dysphasia.

The major differentiating feature of word finding difficulty among dysphasics can be detected only by taking into account the pattern of free conversation. This feature is the availability of lexical terms in relation to the level of verbal fluency—particularly the prevalence of grammatically organized runs of words. The important point to be made here is that conversation and narrative speech must be used as an indispensible part of the dysphasia examination. Highly structured stimulus-response tasks are necessary but not sufficient for evaluating the status of the lexical component of language.

The retrieval of lexical terms is not a simple dimension of dysphasic impairment, since various semantic word categories undergo different fates in different patients (Goodglass *et al.*, 1966). Thus, in the design of a dysphasia examination, it is necessary to explore specifically the availability of object names, action words, numbers, letters, colors and proper nouns.

There is another dimension of variability in lexical disturbance. This one has to do with gaining insight into the way in which the lexical retrieval process has broken down. For example, do we have the grounds for stating that a particular patient recovers a trace of the acoustic structure of the target word but fails only at the step of implementing the corresponding motor speech act? Or can we say about another patient that there is a total dissociation between the lexical representation of a word and the conceptual referent? Or in another case does the patient seem simply to have forgotten some of the infrequent items in his premorbid vocabulary?

Existing dysphasia tests are at best rudimentary in their ability to supply the answers to these questions, although a number of experimental investigations have touched on these problems (Howes, 1967; Barton, Maruszewski and Urrea, 1969; Goodglass *et al.*, 1976). Nevertheless, clinical observation of the qualitative features of patients' behavior suggests that in some cases each of the above mentioned mechanisms is at work. Thus a patient may grope unsuccessfully for a word and, when it is offered by the examiner, he may repeat it dubiously, stating that it does not sound familiar. When this behavior occurs in a context of good auditory comprehension one is forced to entertain the idea of a total dissociation between the referent and its lexical representation. Other qualitative indices come from the presence and type of paraphasic response. A patient who responds to a picture of a baby's rattle with "cradle" would seem to be suffering a different retrieval disorder from one who calls it a "tattle".

It is clear then that the assessment of lexical function offers many challenges to the perceptive examiner. Even though the formal structure of the dysphasia test may not allow for direct probing of the nature of the lexical breakdown, careful clinical observation provides many clues as to what may be going on. Such observations, in conjunction with controlled studies may lead to a behavioral model of how the retrieval and production of words occurs and, in turn, to the developement of revealing examination techniques.

The focus of this discussion has been on the expressive aspect of lexical impairment. The receptive side is equally important. Analogously with the examination of lexical production, the obvious means of testing is to ask the patient to point to or pick up objects as they are named by the examiner. Here, the same variation among semantic categories may be noted as has been cited in the production of words. Numbers and letters may be perfectly understood by patients who cannot identify the most common object. Loss of comprehension of body part names in the presence of good performance with other objects is a common feature in dysphasia. Indeed the examination of body parts is particulary revealing of the multi-layered nature of lexical comprehension. It is commonplace to observe that the patient, given the instruction "show me your eye", immediately brings his hand up to his face only to grope uncertainly around his face. If at that moment he is instructed to shut his eyes, he will do so at once, without being aided in understanding the original request. It appears then that the general postural response to "eyes" may be preserved while the exact lexical reference is lost. At the same time the utilization of the term "eyes" as part of a familiar command has little to do with the processing of the term out-of-context.

There is still another dimension of lexical decoding to be mentioned—albeit one which is totally neglected in the standard examination. We refer to the make up of the semantic field of a word presented for comprehension (Goodglass and Baker, 1976). While a child, hearing the word "engine" may point to the same picture as does an adult, the significance of the word in terms of its complex of associated meaning is clearly different in the two instances. So it has been shown to be the case for dysphasics (Goodglass and Baker, op. cit.). The more severe the functional comprehension deficit, the more impoverished the association network of words that can be recognized as related to a particular target.

c) *Syntactic:* The evaluation of syntax in dysphasia is critical for purposes of diagnosis, yet it has little status in the formal examination for aphasia. The chief reason for this state of affairs is that it is virtually impossible to assess syntax by means of a structured test, depending on right or wrong answers. Such a test inevitably becomes a metalinguistic exercise in which the ability to detach oneself from the speech act in order to apply rules consciously is a central factor. In this type of test one often finds it impossible to distinguish between the patient with syntactically facile conversation and the one who is reduced to little more than a sequence of one or two word sentence fragments.

The underlying consideration is that dysphasia may selectively impair the capacity for generating syntactically organized units, while leaving word finding skills relatively intact. Agrammatism, as this disorder is called, is readily described in terms of the linguistic surface elements which are omitted or retained. For example, articles, prepositions, pronouns, auxiliary verbs, verb and noun inflections and relational pronouns and conjunctions tend to drop out of speech, leaving a predominance of uninflected contentive words—the principal nouns and verbs. However, a description in terms of missing elements does not do justice to the basic disorder. The agrammatic patient is deficient in the realization of the rules by which syntactic relationships are embodied in sentence form. A patient with severe agrammatism performs as though he has no concept of the role of the grammatical morphemes (Andreewsky and Seron, 1975). Yet his appreciation of the underlying semantic value of the intended grammatical construction is unimpaired (Gleason, Goodglass *et al.*, 1975); his fragmented delivery leaves the listener the task of understanding relationships which are normally expressed by means of grammatical words.

At milder levels of agrammatism, simple complete sentences are available, but there is a clearcut restriction in the variety of grammatical forms. The patient tends to revert to simple active declarative sentences as substitutes for more complex forms, often interspersed with fragmented sentences.

As indicated above, it is of dubious value to test the patient by requiring him to perform a transformation on a given sentence, as in converting a present to a future or an active to a passive sentence. By these criteria many patients whose spontaneous speech is perfectly grammatical appear to be defective. They cannot perform under the metalinguistic requirements of this type of task. Another approach to examining the grammatical repertory is to require repetition of sentences incorporating a variety of syntactic demands such as complex verb tenses, subordinating constructions, passives and conditionals. Indeed, this approach is often fruitful, because many patients with agrammatism convert the spoken model into a form which conforms to their spontaneous agrammatic speech as do most children (Brown, 1973). For example given the sentence, "Is the door open" such a patient would say "Door is open?", using intonation only, rather than inverted subject-verb order to denote the interrogative.

The repetition technique has several shortcomings which disguise the patient's underlying abilities. Many patients experience considerable facilitation of speech under conditions of repetition. They succeed in reproducing sentences of a complexity which would be quite impossible in spontaneous production. On the other hand, as noted in the discussion of the clinical forms of dysphasia, there is an equally important group of dysphasics for whom repetition is a particularly disabling task (conduction dysphasia). Because of these problems, the evaluation of grammar has understandably been by-passed in most published dysphasia tests. True, the objective measurement of grammatical production via structured pass-fail test items is as yet an unsolved problem, but this circumstance does not permit us to overlook this vital aspect of language. The technique to be proposed here depends on the elicitation of conversational and narrative speech, which is then rated for the repertory of grammatical forms which are used.

While the notion of agrammatism is historically associated with speech production, the decoding of grammatical structures is an important part of auditory comprehension, which is also generally neglected in most dysphasia tests. It has not been recognized until recently that the apparently good auditory comprehension of Broca's dysphasics is based primarily on their ability to decode the contentive words of incoming messages, to apply word order strategies to the understanding of subject-verb-object relationships and to depend on context and real-world knowledge to extract sentence meaning with a high degree of accuracy. Zurif *et al.* (1972) have shown that when discrimination between two similar sentences depends on the precise use of a grammatical rule, Broca's dysphasics break down in comprehension along lines which suggest a parallel to their expressive disorders. Luria, too, (1970) pointed out that the ability to understand certain grammatically encoded logical relationships is particularly vulnerable in parietal lobe dysphasias. The difficulties may be exemplified by contrasting the two possessive expressions "the dog's trainer" and the "fireman's hat". In the first instance we have a reversible relationship in which the "trainer's dog" and the "dog's trainer" are both possible. The decoding of this expression requires extracting from the minute grammatical morpheme "s" the significance of the relation between the two nouns. The dysphasic who has a deficiency in grammatical decoding, given a picture of a dog and a trainer, is very apt to point to the first of the two nouns which he hears, rather than recognizing that the 's signals that the following noun is the head word. In the case of the 'fireman's hat' the semantic irreversibility of the possessive tends to reduce the ambiguity of the item. The possessive, just illustrated, is one of a large class of logico-grammatical relations (Luria, 1970) which can be tested in a way which clearly distinguishes them from the decoding of the contentive terms. The well known Token Test (De Renzi and Vignolo, 1962) owes its sensitivity to mild dysphasic comprehension disorders, to the fact that it places heavy stress on the comprehension of prepositional relationships—which also belong in this class. However, the Token Test has the disadvantage that logico-grammatical tasks are confounded with sheer quantity of material and with memory demands. In the test procedure to be proposed later in this section, we strive for a more analytic approach—separating vocabulary demands from length of message and from special logicogrammatical demands (*cf.* Goodglass, Gleason and Hyde, 1970).

## Diagnostic Aims of Dysphasia Testing

Diagnosis in dysphasia implies the effort to identify the configuration of deficits in a particular case with one of the dozen or so recognized syndromes. Secondarily, it implies the identification of the probable site of the lesion, on the basis of established correlations between syndromes and their most frequent lesion sites.

Several serious reservations related to localizing the source of dysphasic syndromes must be borne in mind and we have already alluded to these reservations. First, even if one were to assume that the ideal form of each dysphasic syndrome corresponds to a replicable lesion site, the chances that any dysphasic patient will present one of these ideal types are in the neighborhood of 3 in 10 in the case of vascular disease and much less in the case of trauma or tumor. Individual variations in cerebral vascularization, in the availability of collateral circulation, and in the precise location of embolic or thrombotic obstructions make for a wide variation in the actual spatial extent and completeness of cerebral infarction. There is no reason why the site of infarction in a particular case need correspond to the limits of the lesion supposedly identified with a named syndrome. It is of course permissible, however, to identify fragmentary correspondences between observed and ideal syndromes. Such evidence is admissible for making inferences as to the brain sites which are probably implicated—though at a reduced level of certainty.

A further consideration in syndrome localization is the fact that there is considerable variation in language organization from individual to individual. Thus, identical lesions in different patients sometimes produce widely differing symptoms. The most conspicuous basis for individual variation in brain organization is cerebral dominance. Dysphasia in left-handers has been repeatedly observed to deviate from the patterning of expected syndromes (Goodglass and Geschwind, 1976). Similarly, acquired lesions in young children produce very different effects from identical lesions in adults. There is a distinct progression over the years—even into later maturity—as to the way in which the mediation of particular language functions is distributed in the brain. Correspondingly, there is a marked variation in the behavioral effects of identical lesions, as a function of age (Brown and Jaffe, 1975; Obler et al., 1978).

Having emphasized in advance the limitations on the possibility of lesion localization through dysphasia testing, we can consider this goal in its positive aspects. Whether or not a particular dysphasic's performance conforms to a classical diagnostic type, foreknowledge of the possible categories provides the examiner with a set of hypothetical frameworks against which the performance can be judged. Moreover it provides guidelines as to where to probe. Once the similarities to and differences from classical diagnosis have been mapped out, one has a useful shorthand for characterizing the outstanding features of the case at hand. For example, in the case of a patient whose performance straddles the syndromes of Wernicke's aphasia and anomia it is useful to be able to say, for example, that like the anomic, he is fluent in syntax and articulation and manifests a dense word finding difficulty, with blocking and circumlocution in the face of substantives but without press of speech or paraphasia; that like the Wernicke patient, however, he has difficulty in understanding speech and repeats, in bewildered fashion, words spoken to him by the examiner. In most instances, a dysphasic with a mixed picture like the foregoing makes the examiner aware of the fuzzy transitions between the forms of dysphasia associated with lesions in contiguous areas.

The importance of knowing the side features associated with a syndrome cannot be overemphasized, since it guides the examiner as to where to probe and avoids haphazard listing of findings in which special features may have been overlooked. For example, the examiner who detects that a patient can no longer read but can write, speak and understand ought to know that a common cause of dyslexia without dysgraphia (pure word blindness) is an infarction of the left posterior cerebral artery. This may also result in an amnesic state because of the occasional involvement of the medial temporal lobe and hippocampus in the extent of the lesion. He should know that an expressive and receptive dysphasia for visually presented colors is associated with this disorder. His examination will then include a probe for recent memory functions and for color word usage and his description of findings will refer to the presence *or absence* of these possible associated features.

In summary, the diagnostic approach assumes that the best prediction of lesion site in the case of near fit of symptoms with a standard syndrome is that locus most regularly associated with the syndrome. Second, that even partial approximations of a named syndrome provide useful localizing inferences. This approach further provides a shorthand for characterizing the configuration of symptoms. At the same time the examiner must be aware of the pitfalls of a dogmatic approach to the relation between functional impairment and lesion locus.

## The Dysphasia Examination as a Case Study

The implication of the diagnostic approach to testing is one of fitting the patient's configuration of symptoms to a standard of some sort, even though one is fully aware of the abuses when this approach is followed mechanically. These abuses can be summed up in terms of the danger of blindness to or suppression of those behavioral features which may be irrelevant to the syndrome which dominates the thinking of the examiner. The danger can also be put in terms of losing the patient as a unique adapting individual to preoccupation with the classification of symptoms.

To approach the assessment task as a case study means to endeavor to understand not only the nature of the mental processes which are disturbed but the compensatory processes which the patient brings to bear in attempting to cope with the language demands of the test situation. The case study approach means being free to explore and test the limits of the patient's abilities under different conditions. The examiner may be guided by a test protocol and make use of prepared test stimuli. However, he is not constrained by a prescribed procedure and is prepared to improvise new tests to gain further understanding of the patient's change in functioning.

For example, during the initial examination of free conversation, one may note whether the patient spontaneously uses pantomine as a means of demonstrating the concepts which he cannot name. If pantomining is conspicuously present or absent during oral conversation, how does this presence or absence compare to performance of pretended action upon command? One may recall Goldstein's (1948) observation that amnesic dysphasics are often remarkably expressive in pantomine while being unable to name.

As an instance of the adaptation of lower level mechanisms for compensatory purposes we may note whether the patient relies on reciting (silently or aloud) a word series in order to retrieve a term embedded in the series. The most common use of this device is observed in number naming. The dysphasic who cannot name a number (*e.g.* "11") directly may start at "one" and count up to 11 saying that target word with some emphasis. In this instance, the patient recognizes that serial recitation is available to him and he turns to this low level device, which is aided by his ability to recognize the target and to extract it from the series. It is notable that with numbers above a certain level—*e.g.* 20, the patient may not be willing to count. Thus he may end up by calling the symbol "22" "two two", while he has successfully produced the word "11" through counting. While serial recitation obviously applies to number words, it is sometimes used with other terms of a small homogeneous group, such as colors, articles of clothing and denominations of coins. Thus confronted with the color "green" the patient may be heard to whisper "blue, red, yellow, green", then repeat the target word "green".

The foregoing type of performance raises the question as to whether the patient should be credited with a correct response to the stimulus word, when his retrieval is based on serial recitation. On first consideration, the answer is clearly "no", because the mechanism is one of recognition, followed by repetition, rather than one of directly associating the response to the stimulus. On further examination the issue is less clearcut, since the patient may perform serial recitation silently and rapidly and achieve the same result without giving the examiner clear evidence of his roundabout technique. Moreover even normals may use mnemonics of a related type in recovering target words. Our opinion is still that naming via serial recitation does not constitute a correct word retrieval, but that qualitative note should be taken of the patient's capacity to use this technique.

The foregoing illustrations emphasize the analysis in depth of disturbed function which characterizes the case study approach. The goal of understanding the patients as a totality entails the integration of such analyses based upon as complete as possible a sampling of areas of function, both in language and in the closely related intellectual functions of cognition, perception, and memory and on assessment of emotional status and social adaptation. These behavioral observations are related with neurological findings and, if possible across a period of time to trace the evolution of the patient's function as he recovers.

As described here, a complete case study would appear to be a rarely attainable ideal. In practice the examiner, limited in the time he can devote to studying a patient, is forced to be selective as to which abnormalities are examined in depth.

## Quantitative Aspects of Assessment

Whatever the application of the assessment of a given patient it is important that a quantitative measure be provided for each aspect of performance—*e.g.* articulation, grammatical organization, word comprehension, naming on visual confrontation, etc. This means that a large enough sampling of items of each type should be provided to assure that the patient's performance is representative of his capacity at the time of testing. Further, it is important that some means be available for statistically comparing degrees of impairment on various types of tasks, which may be intrinsically very different in difficulty. For example, a score of seven out of ten objects correctly named does not necessarily indicate the same degree of impairment as seven out of ten objects correctly identified by pointing in response to the spoken object name. For most dysphasics matching to the spoken name is easier than independently retrieving the name. Hence, it is necessary to have norms to tell us which scores on each task represent equivalent degrees of impairment. Such normative information is necessary even for diagnostic purposes, since it is the pattern of *relative* impairment among components of linguistic ability which determine the type of dysphasia. The experienced clinician carrying out a bedside examination has, from his experience, internalized a working knowledge of the relative difficulty of various language tasks which he is accustomed to using.

There are applications of dysphasia assessment which make more explicit demand on quantification than is the case in diagnosis. One such application is differentiation between mild dysphasia and non-dysphasic performances, an operation for which norms related to the age and education of the patient are required. Here again the experienced clinician, having at his command informal quantitative standards may make judgments approaching in accuracy those of objective test scores.

Finally, the most rigorous demand on reliable quantification lies in the assessment of change between successive performances, whether the question is one of improvement after an initial insult or deterioration in the case of suspected progressive disease. It is in this application that objective testing based on adequate sampling is usually far more valuable than the clinician's judgment for the reason that smaller differences in performance can be reliably measured.

Among the various dysphasia tests which have been published, some are preferable for diagnosis and case study, others for fine quantification, although the astute examiner can use any one of them with his own adaptations for particular needs.

# 2. Non-Language Aspects of Assessment

Since dysphasia is only one of the many possible physical and psychological by-products of brain injury, the dysphasic patient usually has other deficits than linguistic ones which need to be evaluated. These depend on the nature and extent of the underlying damage and, of course, on the structures which have suffered. In some instances these behavioral changes interact with language and may limit the patient's ability to participate meaningfully in language retraining.

These associated deficits fall into two general categories: those which are non-specific indicators of organic psychological deficit and those which, like dysphasia itself, are selective losses of higher perceptual and motor skills related to specific structures in the brain.

## Non-Specific Indicators

a) *Slowness in understanding and adapting to changing task demands,* expressed in its most severe form in perseveration—the continued tendency to respond in terms of a previous task, in the face of changing task requirements. Closely related to this deficit is the inability to produce new ideas or initiate activity spontaneously i.e. the patient tends to remain inactive or persist in the same activity until stimulated externally.

b) *Reduction in scope of attention* to concurrent aspects of a stimulus—expressed as a response to a simplified construction of the stimulus.

c) *Loss of capacity for abstract thinking*—expressed during testing as an inability to relate concepts to each other through their common properties or class membership, as when required to sort objects into categories or explain the similarity between two terms (*e.g.* "eye" and "ear"); also expressed by the inability to understand the implications of a proverb, beyond the concrete surface meaning of its words.

d) *Stimulus boundness*—the tendency to respond to a stimulus in terms of its immediate sensory attributes or most habitual manipulable aspect, "short-circuiting" more difficult abstract task demands. Expressed in all modalities of testing: verbal, visual motor, etc. in the form of substituting behavior totally determined by familiar stimulus demands for the more complex goal set by the examiner.

e) *Memory disorder*—Some degree of memory impairment is virtually universal as a by-product of brain damage and dysphasia understandably produces the most severe impairment in verbal memory as tested either by the ability to repeat lists, to recognize the recurrence of words in a list (Cermak and Moreines, 1976) or to demonstrate, by sequential pointing, the retention of short sets of objects names (Goodglass *et al.*, 1970). Visual memory is also usually impaired to a milder degree as tested by design reproduction tests (Graham-Kendall, 1946) or Benton's Visual Retention Test (1945) or by multiple choice recognition tests (Benton, 1950).

Memory for the recent past is less commonly disturbed in dysphasia, as these patients normally quickly learn their doctor, their speech therapist, their daily routine and keep track of time. Severe defects of recent memory usually require lesions of the hippocampus or of certain midline limbic system structures which are not generally implicated in dysphasia, yet which may occasionally be present.

f) *Restriction in problem solving strategies related to hemispheric lesion.* Observation of the approach of normals and of right brain and left brain damaged patients to complex drawing or constructional tasks reveals particular deficits characteristic of each of the hemisphere damaged groups, when compared to the normals—deficits which are not readily quantified. Whereas normals readily alternate between attention to global configuration and the filling in of detail, dysphasic patients, having impaired left hemisphere function, tend to orient primarily to the outer configurations and to simplify or omit inner detail. They contrast in performance with right brain damaged subjects who may ignore or violate the configuration of the whole, while focussing on segments or isolated details. These features are particularly well brought out by the drawings of the Rey Osterrieth complex figures (1944) and in the Block Designs of the Wechsler Adult Intelligence scale (E. Kaplan, personal communication).

The assessment of these non-specific cognitive impairments will not be further elaborated in this section as they are not unique to dysphasics and are part of the examination which any brain damaged patient should undergo. For a full treatment of these problems the reader is refered to Lezak (1976) or Goodglass and Kaplan (1979).

## Specific (Localizing) Non-Language Deficits

In contrast to the foregoing non-specific deficits, there is a group of selective disorders of higher perceptual-motor functions, which are analogous to dysphasia in their correlation with lesions of one hemisphere or a particular lobe. These will merely be listed and defined at this point.

a) *Apraxia*—loss of ability to carry out purposeful movement—commonly associated with lesions in or near the language zone of the left hemisphere.

b) *Acalculia*—loss of the ability to carry out elementary arithmetic processes. When this disorder is disproportionately severe in relation to other intellectual impairments it may reflect injury to the dominant parietal lobe. Less profound impairment of calculation ability is common with damage elsewhere in the left hemisphere and does not have further localizing value. Patients with right hemisphere disease show a distinctive problem with written calculation, characterized by confusion in the spatial placement of numbers and the left-to-right procedure in adding or subtracting columns of numbers.

c) *Finger agnosia*—a further by-product of damage to the left angular gyrus (posterior parietal region) may be a restricted impairment of body image—one confined to the fingers of the hand. The patient so afflicted may be unable to match his finger with its name (ring-finger, index) or to match fingers on one hand to those on another hand (patient's or examiner's).

d) *Constructional apraxia*—this term refers to a marked impairment in the ability to construct geometric figures with 2 or 3 dimensional blocks or to draw with pencil on paper, a disorder most commonly found with right parietal lesions. It is distinguished from the simplified, inaccurate drawings of left brain injured patients by the severe breakdown of the spatial framework relating parts to each other.

## 3. Assessment and Therapy

For many speech therapists a major goal of assessment is to derive a prescription for speech therapy. Clearly, one cannot undertake remedial treatment without knowing in some detail the level and quality of the patient's linguistic functioning in each of the modalities of input and output. A number of straightforward consequences for therapy appear as outcomes of particular test configurations. For example, if the patient is found to have dyslexia with dysgraphia with sparing of speech and comprehension the therapist is prepared to work exclusively with written language. If the patient understands well but can produce only stereotyped syllables in his efforts to talk, he may be a candidate for melodic intonation therapy (see section on rehabilitation). If he produces copious unmonitored paraphasia, an early goal of therapy is to slow his speech and train him to monitor his output.

While a knowledge derived from testing of the patient's language profile is a prerequisite for therapy, it is not an automatic prescription for the procedures to be followed. In some instances the therapist may wish to concentrate efforts on the most deficient areas; in others, to train the patient to compensate for absent functions by dwelling more at length on those which are somewhat preserved.

Many therapists take the patient through a period of diagnostic therapy, *i.e.* exploring the avenues through which he is most responsive and shows most promise of improvement. This procedure supplements diagnostic assessment. For some it is the only assessment used.

## 4. The Formal Dysphasia Examination

### a) The Free Conversational Sample

The evaluation of dysphasia should always begin with a sampling of conversational and expository speech. If the examination is a formal quantitative test, this sample should be tape recorded for later analysis. While this part of the examination is only a small fraction, in time, of the total examination, it yields vital information which totally escapes recording in the objective testing of stimulus and response pairs, which constitutes the testing of input and output modalities.

Not only does the pattern of free speech contain diagnostically critical information, but it maintains considerable constancy over the period of recovery. For example, the patient with Wernicke's dysphasia commonly shows marked improvement in auditory comprehension over the first several months post onset. At the same time he continues to speak with the same characteristic pattern marked by frequent paraphasia, a lack of substantive content, and ease of production at the level of phonology and basic grammar. Thus while one traditional hallmark of Wernicke's dysphasia loses its identity in the area of objective scores, the chief features of free discourse retain the same pattern over a longer period.

The choice of a standard means of eliciting extended speech is a difficult problem in test design. The procedure adopted at the Boston VA Hospital Aphasia Center (and incorporated in the Boston Diagnostic Aphasia Exam— Goodglass and Kaplan, 1972) is to begin with simple "small talk", designed to elicit such everyday phrases as "yes", "no", "I don't know", "I hope so" and proceed to a more open ended conversation about the patient's previous work and his current illness. The purpose of the sampling of easy conversational expressions is to give the most impaired patients an opportunity to display their restricted conversational repertory, which usually includes at least some of these formulas. The examiner then leads the patient into a conversation. This conversation is continued for about five minutes, or until the examiner judges that a typical sample has been obtained of the best that the patient can produce. The free conversation is then followed by the presentation of a picture-situation depicting several concurrent and related activities in a kitchen (The Cookie Theft Scene), and the patient is invited to tell all about the picture.

The advantage of a standard picture-based narrative is that it provides a common basis for comparing patients with each other. The picture-narrative, in contrast with free conversation, requires the patient to use a pre-determined vocabulary rather than improvise a conversation with whatever associations come to mind. This combination of open-ended conversation and structured narrative may produce paradoxically different styles of delivery which are very revealing of the patient's speech strategy.

Thus, the opening free conversation is the test procedure which is usually richest in diagnostic information, in relation to the time it takes, aside from its value as a vehicle for establishing a comfortable testing relationship with the patient. There is however a certain amount of interviewing skill required to elicit a respresentative sample of the patient's best effort in conversation, particularly in the case of the patient with sparse, effortful and agrammatic speech. One topic about which a conversation can often be elaborated is the patient's recall of the onset of his illness, how and where it happened, the detailed sequence of events, his description of his condition immediately after onset and his evaluation of his improvement. Other topics which are often productive include eliciting an account of the patient's work and hobbies, his military service, how he spends vacations, or whether he has children and what they do. It is well to avoid probing for one-word factual answers; this often leads to guessing by the examiner and yes-no responses by the patient. The examiner should participate in a way that encourages some extended speech by the patient. On the other hand, some patients are totally unable to initiate an utterance in response to a very open-ended question, such as "Tell me more about it." Thus, the proper interviewing technique is one which facilitates speech with encouragement and leading questions which are neither too unstructured nor too constraining. In the case of fluent dysphasics who may be disinhibited and garrulous, interviewing technique may be unimportant; in the case of the very severe Broca's or global dysphasic no amount of assistance may elicit any speech. In the latter instance, of course, the attempt to interview should be brought to a close fairly promptly with a remark acknowledging the patient's hardship in expression.

*Analysis of free conversation*—The first level of analysis of the conversational sample involves a decision as to whether the speech pattern can readily be assigned to the fluent or non-fluent category. Secondarily, if the pattern is fluent, the sample should reveal whether there is significant use of paraphasia (as in Wernicke's, conduction or, transcortical sensory dysphasia) or not (as in anomic dysphasia).

The task of the clinician is to learn to tune his ear to certain significant dimensions of the patient's speech output pattern. As a guide to this skill, we reproduce the scales from the Profile of Speech Characteristics of the Boston Diagnostic Aphasia Test (Goodglass and Kaplan, 1972). (Fig. 2.)

The dimensions scaled in this profile are those which are vital elements in differential diagnosis, for which there is as yet no more objective means of obtaining a score.

The first four "Melodic Line", "Phrase Length", "Articulatory Agility" and "Grammatical Form" tend to be associated in that their impairment is characteristic of the nonfluent (Broca type) dysphasia, while their preservation is the mark of the fluent (Wernicke, anomia) or "posterior" dysphasias. In spite of this correlation, they may vary autonomously from each other; hence the examiner should learn to listen to each of the aspects by itself.

*Melodic line (Prosody)*—This refers to the normal intonational pattern of a sentence, with its pattern of stress and rising and falling pitch. Melodic line is almost invariably disrupted when speech is uttered haltingly in short word groups, typically of 1 to 3 words. Such short word clusters are characteristic of many Broca's dysphasics, who find it impossible to retrieve sentence patterns and therefore speak in short agrammatic groupings. In these subjects, it is the syntactic disability and reduced access to vocabulary which disrupt prosody. Articulation *per se* may not be damaged—or at least not damaged perceptibly in conversation, although it usually proves to be defective when stressed by sentence repetition or by difficult test words offered by the examiner. Such patients commonly have a repertory of short, stereotyped sentence patterns (*e.g.,* "I know what it is but I can't say it"), which may be emitted with normal intonation. These should not be the basis for assigning a normal rating in melodic line.

Patients who do have labored and awkward articulation are rarely able to maintain a melodic contour across a sentence. Each word may be produced with so much effort that the normal stress pattern is set aside.

The most severe ratings of disturbed melodic line are reserved for patients who impose a rising and falling sentence-like intonation on each individual word or short word grouping. Typically these patients are also both agrammatic and impaired in articulation. With some patients, the peaks and downswings are markedly exaggerated, giving their delivery a childish quality—each word in its own sing-song. Intermediate ratings are assigned for those subjects who produce some short (3—4) word sentences with normal prosody, but in a context of an otherwise halting and dysprosodic delivery. Normal prosody, of course, is quite compatible with incomplete sentences and incorrect grammar. However, prosody should be considered normal only if sentences (or incomplete sentence fragments) include uninterrupted strings of five or more words. The reason for this seemingly arbitrary rule is that non-fluent patients frequently recover to the stage of producing occasional short, well intonated sentences though a background of halting speech is still evident. If the rating of "normal" prosody were liberally applied to these cases, the scale

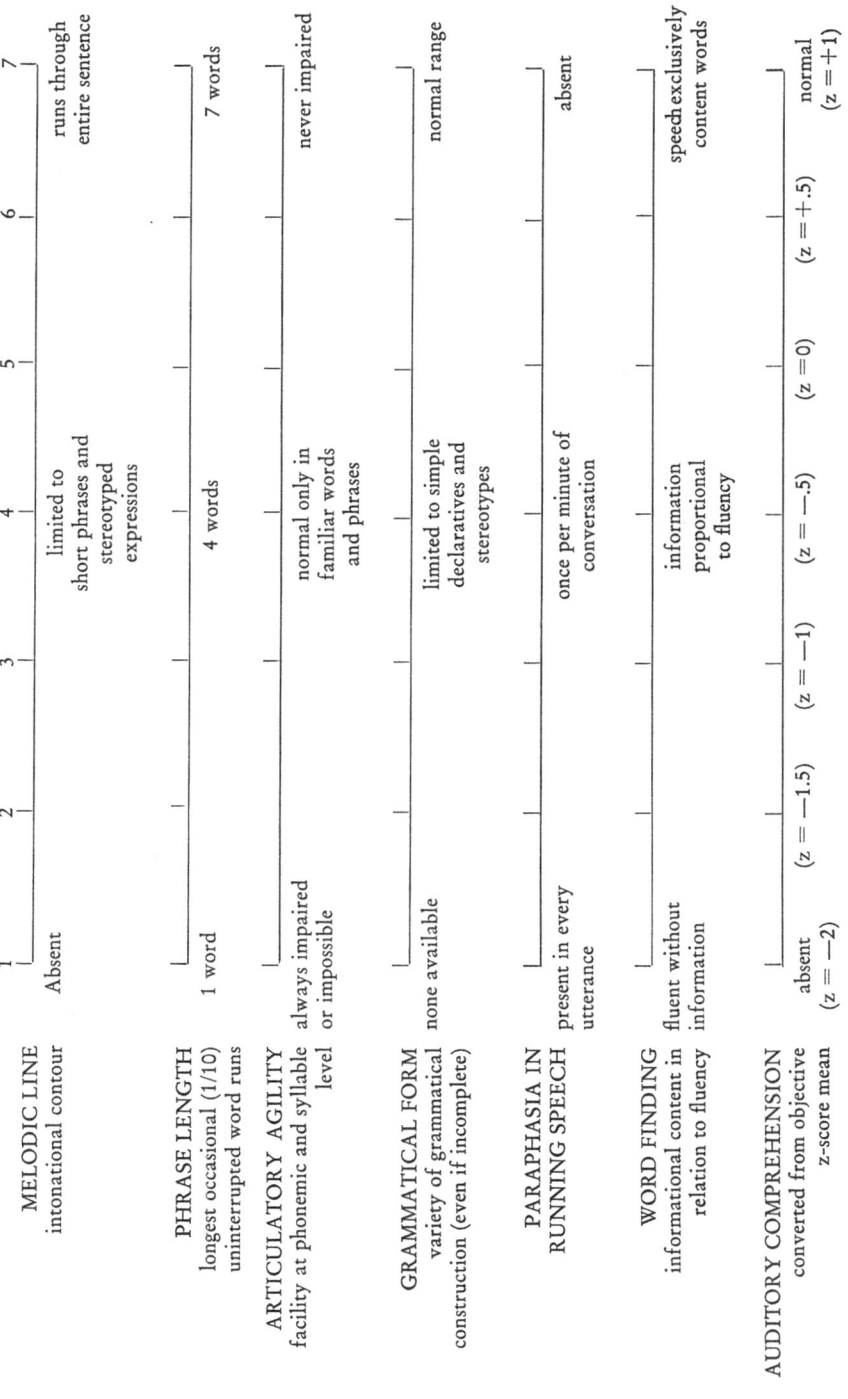

Fig. 2. Rating scale profile of speech characteristics

would lose its ability to discriminate between basically fluent and basically non-fluent dysphasics.

*Phrase length* (runs of uninterrupted words)—In an effort to find the single feature which most consistently corresponded to the clinical impression of speech fluency, Goodglass, Quadfasel and Timberlake (1964) decided on a measure based on the frequency distribution of uninterrupted word-runs of varying length—from one word to seven or more. It is clear that this measure must correspond very closely to a count of words per minute, which has been used by various investigators (*e.g.*, Howes and Geschwind, 1964). We find the phrase-length count more useful on two grounds. First it captures those patients whose speech is interrupted by frequent word finding blocks, but who nevertheless have occasional long uninterrupted runs, although their rate across a full minute may be no better than that of the non-fluent halting Broca's dysphasic who keeps up a steady output. The second advantage of this measure is that it can be estimated with adequate reliability while listening to the patient talk and does not require counting words for several one minute samples with a stopwatch.

"Phrase length", in our usage, does not refer to grammatical phrases, but merely to the number of words uttered between pauses. The "phrase length rating" is the longest occasional uninterrupted word run. By "occasional" is meant one which may be expected to occur in about 1/10 of the patient's starts. That is, the fluent (Wernicke, conduction or anomic) patient may initiate many utterances which are terminated by word-finding failure after 1, 2, or 3 words. Interspersed among these aborted attempts, however, are strings which continue for 6 or 7 words or longer. Thus the characteristic of the fluent dysphasic is the preservation of the potential for occasional word strings beyond the 4 or 5 word upper limit which marks the non-fluent (Broca type) speech output. The fact that an utterance may be filled with paraphasia or make no grammatical sense does not interfere with counting it towards a phrase length rating.

While the phrase length rating is a useful indicator for clinical diagnosis, it cannot be applied mechanically. The most important interacting factor is that of severity. As a Broca's dysphasic recovers, he may reach a point where occasional sentences of six or more words are delivered without a break, although other utterances still betray the agrammatism and articulatory awkwardness that typify this syndrome. Conversely, some "fluent" dysphasics are so severely limited in strategies for evading their word finding disability that they never attain a long run of words. The indices of their fluency lie in the articulatory ease and natural intonation of their fragmentary utterances, as well as in their grammatical structure.

As noted earlier, in connection with speech melody, stereotyped expressions ("I know what it is but I can't say it") should be excluded from consideration in rating phrase length.

*Articulatory agility*—Along with phrase-length, facility of articulation or its loss are the central features in deciding whether an dysphasic speech pattern is of the fluent or non-fluent variety. Articulation tests yielding an objective score are available for dysarthrias due to disorders of muscular control of the speech apparatus or poor speech habits of childhood. These consist of lists of

words in which each English sound appears in word initial, medial, or final position. Such tests are of little value for dysphasia. The dysphasic patients whose
articulation is defective in free conversation (Broca's dysphasics) commonly
do much better when given a single word for repetition. Moreover, their defect
may not involve the substitution of a transcribable error, but may consist
simply in laborious or awkward (*e.g.*, over-aspirated or prolonged) delivery.
On the other hand, the patients whose articulation is most facile (Wernicke's)
may make grossly paraphasic substitutions, which must be counted as failures in
a repetition task, causing the score to give a totally fallacious impression of
articulatory ability.

These considerations led us to the adoption of the rating scale approach
for evaluating articulation. At one extreme (Rating 1) are those dysphasics for
whom speech sounds are always laborious, distorted, or impossible to produce;
at the other (Rating 7) are those whose articulation always sounds normally
agile. The midpoint represents patients whose articulation is normally agile
and accurate for occasional short phrases and familiar words but otherwise
laborious and distorted.

While the foregoing implies that the criterion for scaling is the constancy
of articulatory difficulty or failure, the examiner cannot help but take into
account the degree of disorder, which may not correspond to its constancy.
That is, a patient with slight but constant awkwardness in articulation may
seem intuitively less impaired than one with intermittent articulatory failures
which produce a severely distorted output. We see no resolution for this
problem except to allow the examiner to use his judgment in trading off
constancy against severity to arrive at an intermediate rating.

A perennial problem in arriving at replicable ratings of articulation is the
decision as to whether a misarticulation is a distorted version of a correctly
chosen target phoneme or a well articulated version of an incorrectly targeted
sound. That is, are we dealing with an articulatory error (impairment at the
phonetic level) or a *literal paraphasia* (impairment at the phonemic level).
At the extremes, there is no difficulty deciding. For example, when the patient,
attempting the word "chair" produces "tssair", it is clear that he had the
correct phonemic target in mind but produced a phonetically simplified form
which could be transcribed using a different symbol from the target "ch". On
the other hand, when a patient names "scissors" as "a pair of klizzors" he can
hardly be deemed to have aimed at "s" and produced "kl" because of articulatory problems. This is clearly an instance of phonemic or "literal" paraphasia,—the (usually) facile emission of an incorrectly targeted phoneme. The
decision is easy this case because the substitution "kl" is so remote phonologically from the target "s" that it would be hard to conceive it as an articulatory distortion. One bit of discriminating evidence which is hard to convey
on the written page is the ease with which the erroneous sound is emitted. When
the patient characteristically has effortful and distorted articulation and when
the error in question is produced in labored fashion, it should be treated as
an articulatory distortion—*not* a paraphasia—even though it may be transcribable with another English sound. On the other hand, if the patient
characteristically has runs of rapid and facile articulation of speech and, in this

context, produces a word with a phonemic substitution, transposition or intrusion, such an error would generally be a "literal" (or phonemic) paraphasia.

*Variety of grammatical forms*—The fourth of the scales closely correlated with "fluency" in dysphasia is that reflecting the repertory of grammatical forms available to the patient in his conversation.

At the most impaired extreme are those patients—usually Broca's dysphasics—who have *agrammatism*—a near total absence of the "small grammatical words" of the language. Indeed agrammatism is often described as a dropping out of speech of the customary articles, pronouns, noun and verb inflections, auxiliaries. Closer inspection of agrammatic speech suggests that this style has a more complex explanation than a mere dropping out of grammatical elements. In fact there appears to be a basic loss of the concept of words as having a functional role in a sentence. The severe agrammatic uses words as disconnected, nominalized ideas, which can be placed contiguously without any expressed grammatical connection between them.

At the other extreme, on this continuum, are patients whose grammar shows a normal variety of syntactic forms, with no tendency to omit grammatical function words or inflectional forms. If the examiner hears sentences that begin with subordinate clauses, or observes the free use of auxiliary verbs, nouns or adverbs derived from adjectives by the appropriate suffix, this is sufficient basis for assigning a normal or near-normal rating on this scale. The fact that sentences are left incomplete or may be semantically anomolous need not preclude a high rating for variety of grammatical forms.

At the midpoint on this scale are those patients who produce short, but grammatically complete sentences, restricted in form to subject-verb-object construction, in the present or simple past tense. These are usually patients who have recovered from a more severely agrammatic level. When short complete sentences are scattered in an output which includes grammatically defective forms (*e.g.,* with omitted grammatical morphemes), it is appropriate to give an intermediate rating, such as "2" or "3" on the seven point scale.

*Paraphasia*—Rating the presence and frequency of paraphasia, as well as its quality, is of prime importance in using the free conversational sample as a basis for a diagnostic judgment. First, it should be noted that this scale is set up to rate the incidence of paraphasia *in running speech*. That is, it is intended to discriminate only among the varieties of posterior dysphasia since only they, the fluent dysphasics, can be considered as having "running speech". Specifically this means that the non-fluent dysphasic, even though he may commit some verbal paraphasias is assigned a rating of '7' on this scale. The reason for this scoring convention is that the production of verbal paraphasia on isolated contentive words occurs among all dysphasics; while more frequent in Wernicke's dysphasia, this symptom alone has little diagnostic discriminating power. It is the presence of uncorrected paraphasia—usually multiword paraphasic strings in a context of fluent output which is discriminating. It distinguishes the Wernicke's dysphasic from the anomic patient.

The steps in this scale reflect the frequency of paraphasia during free communication. At the left hand end, the rating of "1" applies to patients who produce a paraphasic error in every utterance. At the midpoint, with a rating of "4" are those who produce a paraphasic error about once per minute of conversation. The right hand extreme of the scale applies to those whose speech is free of paraphasia or to non-fluent dysphasics who produce no running speech.

*Word finding*—Word finding, for purposes of this scale, is a relative, not an absolute rating. It is an estimate of the balance between contentive words and grammatical filler words. By "contentives" we mean nouns, principal verbs, adjectives and adverbs. Unlike the other scales the optimal point is at the middle rating of "4" rather than the right hand extreme of "7". The two extremes represent opposite forms of deviance; one with a very low ratio (approaching ∞) of contentives to filler material; the other with a very high ratio of contentives to grammatical filler words.

This scale (save for the extremes) does not necessarily reflect the severity or mildness of dysphasia. Thus, while a rating of "4" is to be expected from a normal speaker, it can also be obtained by a moderately severe dysphasic, who produces only simple sentences, which include subject nouns, verbs, and objects, along with the appropriate grammatical background.

*Auditory comprehension*—While this variable appears as a scale on the Profile of Speech characteristics, it is placed there only because the overall profile is enhanced as a differential diagnostic tool by the inclusion of a value for auditory comprehension. Auditory comprehension, unlike the other six variables is adequately scored by objective pass-fail items. Therefore the value to be indicated on the profile is a conversion of the average score on the auditory comprehension subtests of the Boston Diagnostic Aphasia Examination (Goodglass and Kaplan, 1972).

*Applying the Rating Scale Profile to Diagnostic Classification*

In many instances (30—40 percent of unselected cases), inspection of the profile of speech characteristics leads directly to a diagnostic assignment. For example:

*Broca's dysphasia:* The first four scales (which reflect various aspects of fluency) will have low ratings, while the last three ratings will approach the right side of the scale.

*Wernicke's dysphasia:* The first four scales will approach normal ratings, while the scales for paraphasia, word-finding, and auditory comprehension will be rated to the left of center.

*Conduction dysphasia:* The profile will resemble that of Wernicke's dysphasia, except for the auditory comprehension scale which will approach a normal rating. It must be recalled that the distinctive feature in this disorder is impaired repetition for which no scale is provided in the profile, since repetition can be adequately scored by objective measures.

*Transcortical sensory dysphasia:* This disorder will present a profile identical to that of Wernicke's dysphasia. It is distinguished from Wernicke's by the remarkable preservation of repetition which is not represented by a scale on the profile.

*Anomic dysphasia:* This disorder will show a profile with ratings approaching normal on all scales except for word finding (informational content), on which the rating will fall to the left of the midpoint.

*Transcortical motor dysphasia:* In its most classic form, the profile for this disorder resembles that of Broca's dysphasia, except for a normal rating in facility of articulation. As in the case of conduction and transcortical sensory dysphasia, the critical variable of repetition is not represented. In this instance, repetition is remarkably intact.

## b) The Evaluation of Auditory Comprehension

Traditionally, auditory comprehension has been treated as though it were a unidimensional capacity, varying only with the length and vocabulary level of the message. Typical of this approach is the classical 'three papers' test of Pierre Marie (as described by Weisenburg and McBride, 1935). This consisted of giving the patient three pieces of paper torn to different sizes with the instructions to "throw the small one on the floor, put the middle sized piece in your pocket, and give me the third piece". In current clinical practice, the ability to carry out commands, graded as to the number of components is still often treated as the primary technique for assessing auditory comprehension. In our experience it is an important portion of the examination but only in the context of other procedures.

We approach this area at five levels: 1) the ability to understand personally relevant material; 2) the comprehension of individual vocabulary items of graded difficulty; 3) the amount of information that can be apprehended in a single message; 4) the ability to grasp and make simple inferences on hearing material unrelated to the immediate situation; 5) the ability to grasp various relational syntactic constructions. In fact, however, the examination of auditory comprehension cannot be reduced to five or any number of pigeon-holes. Each level gives rise to variations which imply additional dimensions.

*Personal relevance*—The patient with an auditory comprehension deficit usually responds differently to questions or commands which relate to his immediate life situation, as opposed to questions concerned with more remote factual material. This disparity may be very dramatic in some individuals. The opening free conversation, in addition to providing a speech sample, affords an opportunity to observe the response to question about the patient's current complaints, his current daily routine, people with whom he is in contact from day to day, such as his therapist, family members, physician. The examiner should be aware of when he changes from one subject to another since the patient's inability to follow a change in topic may lead to the misapprehension that the fault lies with the difficulty of the subject matter.

The examiner of the severely impaired patient should have in his repertory a number of commands which appear to tap a level of comprehension very close to the patient. One of these for patients who wear glasses is "take off (or put on) your glasses". Another is "shut your eyes".

*Axial commands*—In the examination of the patient with a dense comprehension defects it is important to explore the response to a category of commands which refers to postural adjustments of the body axis, which are frequently remarkably well preserved, in contrast to even more elementary commands involving the limbs. Geschwind (1965) suggests that comprehension of whole body commands is mediated by bilateral pathways. A graded series of "axial" commands is suggested here, some of which are suitable for patients in a wheelchair:

    look up
    look down
    look behind you
    lean forward
    stand up
    turn around
    walk forward
    march in place
    stand at attention
    stand like a boxer (or a golfer)
    take a bow.

These can be combined into multiple commands to test the upper limits of this category of item, *e.g.* "stand up, take a step backwards, turn around twice and sit down". It is notable that certain limb positions (*e.g.*, those involved in a boxer's or a golfer's stance) may be carried correctly as part of a grossly axial posture, when pure limb commands (*e.g.* "salute like a soldier") cannot be done. It has been suggested that the command "shut your eyes" owes its resistance to dysphasia to its status as an axial command.

*Testing of comprehension vocabulary*—The ability to understand the meaning of words spoken out of context constitutes an important component of the dysphasia examination and one which varies considerably in difficulty as a function of the examining conditions. In the first place, the examiner must be aware that a word out of context can be much more difficult than the same word in a familiar sentence. For example, the patient who instantly reaches for his glasses when asked to put them on may be unable to point to "glasses" as one of an assortment of 4 or 5 objects on the table. The patient who cannot point to his "eyes" on command may, while searching for them, instantly obey the request "shut your eyes" even as he continues groping for "eyes" without realizing the identity of the target of his search with the word in the sentence.

Obviously, a rare word (*e.g.* "Zodiac") is less commonly understood than a common word like "bed" but many conditions other than frequency may be varied. Is the object named to be pointed to in its natural location in the room or as an arbitrary array set before the patient? Is the selection to be made from a set of a single category (*e.g.*, all objects, all body parts, all colors) or are various categories presented together? Is the choice one out of two or one out of ten possibilities in an array? Dramatic and unexpected differences are often to be found among various semantic categories: letters, colors, actions, objects, etc. (Goodglass *et al.* 1966).

While the examiner is unlikely to test the patient under every possible stimulus condition, he should always be sensitive to the possibility that his results are condition-bound and that a slight change may bring a surprising change in patient performance.

It is recommended that the patient first be asked to point to various objects around the room—ceiling, floor, window, door, chair, pillow, mattress, washstand, mirror, blackboard, as they may be available. The ability to respond to the individual words should be compared to response to a descriptive phrase—*e.g.* "What do you sit on", "Where is the entry into this room", "Where would you see your own reflection", and so forth. The patient should then be tested with an assortment of objects placed on the table before him—*e.g.*, pencil, scissors, spool, comb, knife. The number of objects should not exceed six, lest auditory comprehension be confused with the ability to search and find an object in a large array.

The ability to understand the names of letters, of numbers, of colors and of body parts should be examined with a similar multiple choice procedure. It is common to find that Wernicke's dysphasics fail badly with objects, yet perform flawlessly with letters and numbers. The ability to point to various body parts on oral request is often more severely impaired than object identification.

When the patient fails to understand a body part name, it is worthwhile to present the same word to him in writing. It is common to find graphic input for single words better retained than auditory input—and this is especially true for body part names.

The Boston Diagnostic Aphasia Examination (Goodglass and Kaplan, 1972) provides an interesting variant in the testing of single word comprehension across semantic word types. One test card contains three groups of items, *eg.* 6 pictures of objects, 6 letters, and 6 geometric forms. A second card shows action pictures, numbers, and colors. By testing words from different categories in rotation, one may discover that the patient immediately searches in the group of choices corresponding to the correct target, but that he never succeeds in identifying the correct item. For example, he may look among the geometric forms in the response to the stimulus-word "triangle", but may finally select out the circle. This performance is very analogous to the common phenomenon of pointing to a wrong body part, but one which is in the same body zone as the named target. These behaviors show that the meaning of a word is simultaneously processed at several different levels. A patient may appreciate the connotative or affective component of word meaning but be unable to demonstrate a one-to-one match.

*Pointing span*—While their ability to point to a variety of objects may be unimpaired, most dysphasics are markedly restricted in ability to point in succession to two or more of a group of objects placed before them, maintaining the order of the spoken instructions (Albert, 1972). Goodglass, Gleason and Hyde (1970) found Broca's dysphasics to be the most deficient, when account is taken of their excellent comprehension of individual words. Most of these patients cannot reproduce a sequence longer than two and many fail with a two word series.

In an informal test situation, the procedure is to lay out an assortment of common objects and verify that the patient can correctly point to each in response to its name. He is then instructed to point to two in the order that they are named. The examiner gathers the objects with his hands and says "show me the ... then the ...", immediately aligning the objects in a new order. It is considerably easier for the patient to point in succession when the objects are displayed in position while they are being named, since he can mentally tag the sequence of locations. As a formal test, it is more convenient to make a set of cards with pictures of eight common objects. The cards used at the Boston Veterans Administration Hospital show a cat, lamp, clock, boat, fork, slice of bread, chair, and pen. Four cards are used with the objects in different positions on each card. The words are named while a card is face down and it is then immediately turned up for the subject to point. Different cards are used in successive items, to prevent the subject from associating an object with a particular position.

It is notable that the demands of this test are much more severe than for pointing to a succession of objects in the room. For example, a patient may succeed well in response to "show me your pillow, then the washstand, then the ceiling, then the window", but fail to point to two objects in the pointing span test. The former test may facilitate response by allowing the patient to code each element as to location in the room while the series is being named. The fact that they are in their natural locations may help. This factor is of course absent in the case of an arbitrarily sequenced group of objects.

*Following commands*—Oral commands serve readily as the vehicle for testing the number of informational elements with which a patient can deal in a single message. They can be as short as "make a *fist*" or as long as "tap *each shoulder twice* with *two fingers*, keeping your *eyes shut*". The first contains one informational element; the second, five. The underlined words are considered to be those which carry the independent non-redundant ideas in the message. The above instance illustrates an information-rich message in the form of a single command. Another approach is to complicate the command by stringing together longer series of independent actions (*cf.* Marie's Three Papers Test). The popular Token Test (De Renzi and Vignolo, 1962) adds complexity in the early items by requesting increasing numbers of plastic tokens in which color and form are specified. Later items are made still more difficult by including relational prepositions.

Unless they are constructed with great care, tests of increasingly information-rich commands are likely to vary along several dimensions at once—length, number of contentive terms to be decoded, and syntactic complexity. While they are successful as a practical, wide range measure of auditory comprehension, they are difficult to analyze. This is the case with the Token Test, a popular and sensitive measure of comprehension difficulties in dysphasia. Whitaker and Noll (1972) have published an analysis of the syntactic demands of this test.

*Real life commands*—From time to time a patient is seen with paradoxical contradictions in his response to auditory comprehension. Unresponsive or erratically responsive to object names, he may carry out no commands involving an array of test objects or tokens. He may however perform dramati-

3*

cally well when required to carry out a life-like errand involving bodily action, such as, "Please go over to the windowsill and get me the magazine from under the flower pot." One such patient in our clinic responded promptly and correctly to the request to "Turn the lights off and back on again", when he had been bewildered by much simpler tasks, to be responded to by pointing. The mechanism for this dissociation is obscure—it appears to involve ambulation and real-life practicality, yet even these dimensions have to be explored.

*Response to questions of fact*—One occasionally finds a dissociation in which the dysphasic patient understands questions and commands referring to himself and to objects present before him, but where he fails to grasp questions of fact about things or events which are unrelated to what he sees. In the Boston Diagnostic Aphasia Examination we devote a subtest to a series of items asking yes-no questions of fact about common phenomena and questions based on comprehensions of short paragraphs which are first read aloud by the examiner. The easiest item is "Is a hammer good for cutting wood?" The most difficult is based on the comprehension of a paragraph describing the role of instinct and training in the development of hunting skills by lion clubs.

## Special Tests of Syntactic Decoding

The assessment of auditory comprehension is not complete without specific probing for the ability to decode various syntactic constructions, including the use of prepositions of time and space relations as well as prepositions determining the case of an object noun. Many dysphasic patients who appear to have nearly intact comprehension are in fact deficient in understanding grammatical relationship. Suggestions for testing some of these grammatical elements are as follows:

*Passive subject—object order:* The examiner says "A lion and a tiger had a fight and the tiger was killed by the lion. Which animal is dead?" If the patient cannot speak, examiner asks—"Was it the lion?" "Was it the tiger?"

A second example:

"Mary and Bill got into an argument and Mary was slapped by Bill. Which one did the slapping?" ("Was it the boy . . . was it the girl?")

*Possessive:* A particular demand is placed on the comprehension of the possessive marker 's when both nouns are animate, so that the relationship between the two is semantically reversible. One type of item uses kinship terms of opposite gender as the following:

The examiner says "Suppose we see someone walking across the street there and I say to you, 'That's my sister's husband'. Am I pointing to a man or a woman?" Other example may be "My brother's wife", "My daughter's husband", etc.

With the help of suitable drawings the patient may be asked to distinguish between "the kitten's mother" and "the mother's kitten", "the dog's trainer" vs "the trainer's dog" and "the ship's captain" vs "the captain's ship".

*With/to agency*—When a dysphasic patient is shown a fork and a pencil on a table before him and asked to "Touch the pencil with the fork", in most cases his response is to pick up the pencil and to touch the fork with it. This type of behavior is found even when the patient is shown by demonstration and explanation what he is expected to do. In contrast, when the command is rephrased "With the fork touch the pencil" there is usually a prompt and correct response. Analysis of this process suggests that the dysphasic misuses word order so as to interpret the first named object always as the thing to be picked up and used as the agent of the touching action. When the grammar of the spoken sentence follows this rule he is right; when it is the reverse, he is wrong.

In the administration of this test a number of elongated objects; such as spoon, fork, pencil, comb, stick are kept on hand and used in a random sequence as the two test objects. In order to lead the patient into the proper set for the task, it is advisable to begin by showing two objects (*e.g.*, pen, comb) and saying "Pick up the pen ... Now, with the pen touch the comb". If necessary, demonstrate at this point, then replace the objects and say "Now, touch the pen with the comb." From this point, begin to substitute other objects and alternate randomly between the two forms of the command until it is apparent that the patient

    a) understands the significance of "with", making few misses

    b) tends to follow the principle of using the first named object as the agent

    c) performs randomly.

*Prepositions of time and place*—When we say "The train arrived before the bus" we depend on the word "before" to convey a complex relationship of time between "train" and "bus". This relationship could be conveyed more clearly, but at greater expense in number of words by expressing it in the more redundant form, "The bus arrived first, and then the train." Our experience has been that the terms "to the right of", "to the left of", "behind" are harder for dysphasics than "on", "under", or "in back of".

The following examples are offered to illustrate the approach to testing the comprehension of relational terms.

    *Before/after:*    Do you usually put your shoes on before your socks?

                    Do you put your socks on before your shoes?

                    Do you eat lunch before you eat supper?

                    If I told you the train arrived in town after the bus,

                    which was first, the train or the bus?

                    Does Friday come after Saturday?

                    Does Friday come before Saturday?

*Directional prepositions:* Place a coin (or similar small object) and a toy automobile before the patient and say:

                    "Put the coin behind the car"

                    "Put the coin under the car"

                    "Put the coin in front of the car"

                    "Put the coin to the right of (to the left of) the car".

(The car should face away from the patient during the right-left items, so that "right" and "left" coincide for the car and the patient.)

## c) The Assessment of Productive Speech

The most important single procedure for the assessment of speech output is the sampling of free conversation and narrative speech, described earlier, which should have been done as the opening procedure in examination. The procedures which are described below—testing automatized speech, repetition, and naming are also vital for completing the study of the patient's language function and, in many instances, necessary for a diagnostic assignment.

*Testing oral motility.* The motor control of the oral apparatus is obviously closely related to speech, and if it is impaired by paralysis or ataxia, the difficulty will inevitably be reflected in articulation, resulting in a dysarthria. Still it is possible to observe severe praxic disturbances of oral and facial motility which leave motor speech production unscathed. Conversely, non-speech oral movements may be quite intact in the presence of a profound motor speech impairment. It is important to include an evaluation of non-speech oral movements to see which of the foregoing situations applies. The procedures suggested involve rapid alternating movements which make it easier to detect the presence of minor impairments. A list of movements with scoring standards is included in the Boston Diagnostic Aphasia Examination (Goodglass and Kaplan, 1972). They include pursing and releasing lips, retracting and releasing corners of the mouth, swinging tongue from one corner of the mouth to the other, protruding and retracting tongue, and moving tongue tip up and down from roof to floor of the mouth. These are all done after demonstration by the examiner. In a later section, dealing with the examination for apraxia, testing of some additional oral and respiratory acts will be described.

*Oral agility.* Facility in articulation has already been rated on the basis of the opening conversational sample and the sample of free narrative from picture description. Another approach, and one which can yield an objective score, is to have the patient repeat a single test word over and over at a fairly rapid pace. With this approach, patients who tend to be paraphasic can be aided until they are started on a correct repetition sequence, and the number of error free repetitions can be counted to yield a score. The practice at the Boston VA Hospital (based on the Boston Diagnostic Aphasia Test) is to start with very easy sequences (mama ... mama etc., fifty-fifty ... fifty-fifty) and proceed to longer more complex words (baseball player—baseball player). This technique places no demand on repetition memory span nor on free speech production and allows the patient to display his capacity for articulatory agility in an extended run of repetitions of the same word.

*Automatized speech.* The examination of memorized sequences provides an important diagnostic clue in dysphasia. A very marked discrepancy in the form of severe incapacity in free or conversational speech, contrasting with near normal performance in serial speech, recitation of memorized passages, and singing words with music, forms part of the syndrome of the transcortical dysphasias, along with excellent response to repetition. Unless there is a very marked advantage, the superiority of serial speech as an output mode has little significance. As a performance which is lacking in propositional quality, memorized and, particularly, serial speech is often relatively well retained in severe dysphasics of all types, except conduction dysphasics. However this ability in patients other than transcortical rarely goes beyond counting, saying the days of the week, and partial retention of the alphabet. The items usually included in this subtest are:

a) the days of the week
b) the alphabet
c) counting up to 20
d) reciting the months of the year
e) recall of familiar nursery rhymes
f) recall of words to familiar songs (*e.g.* America, Jingle Bells)
g) recitation of a well memorized passage such as the Lord's Prayer.

The recitation of the words to songs may be tried with and without singing, to see if the melody facilitates the recall of the words. When the patient cannot initiate one of the foregoing responses it is helpful to provide the opening word or phrase. It is common for patients to perseverate an earlier series when asked to recite a new one. In particular, when counting is tested first, many patients begin to count again when they are asked to recite the alphabet. For this reason it is best to delay counting until some of the other tasks are tested.

It is well to note whether patients who recite extremely well respond especially well to cueing with the opening sound in the testing of word naming. These two performances commonly go hand in hand.

## d) Repetition

The function of repetition serves as an important sign of the state of the language apparatus. At the extremes, selective preservation of repetition is indicative of the transcortical dysphasias and anomia; its selective impairment is indicative of conduction dysphasia. A dysphasic patient's ability to repeat depends not only on the length and phonological complexity of the target utterance but, in the case of isolated concepts, on their semantic category and on their status as real vs nonsense words; in the case of phrases and sentences, on their grammatical and lexical composition. The systematic variation of these factors will be reviewed.

The examiner's strategy in exploring repetition will vary, depending on whether he is following a formal testing procedure or a bedside examination. In the latter case, he will enter the repetition task at a level of difficulty commensurate with his impression of the patient's ability and back track to more elementary tasks if he proves to have overestimated what the patient can do.

*Phonology and semantic class.* The formal survey of repetition begins with high frequency one-word units. Starting with conversational expletives we see if the patient can repeat "yes", "no", "hello", "thank you", gradually lengthening to such formulas as "have a seat", "thank you very much" etc. Easy nouns such as "house", "chair", "window" are within the capacity of patients who do not have profound motor articulatory defect, including mild to moderate conduction dysphasics. Phonological complexity is introduced by varying the sheer length of words—e.g. "elevator", "television", but it is complicated by introducing compound words (e.g., butterfly, windowsill) and much more so by adding alliterative features, as in "baseball player", "Methodist episcopal", where the recurrence of bilabial consonants present a difficult challenge, particularly to conduction dysphasics.

The evaluation of repetition errors is critical since patients of different types have different patterns of failure, and a mere noting of success or failure will omit important information. Thus the patient with the speech pattern of Broca's dysphasia usually shows a slight facilitation of speech output in repetition as compared to that of conversational speech, but, in general, his repetition shows a similar quality of stiffness, phonetic distortion, or, in severe cases, complete inability to find the articulatory position for reproducing the target pattern. In contrast, the conduction dysphasic may be worsened in repetition as compared to free conversation, his defect becoming most apparent in the effort to repeat alliterative compound words. While anticipating and awareness of productive difficulty often cause blocking and phonetic distortion in conduction dysphasics, their errors more characteristically show transpositions of sounds and intrusions of some unintended, yet well articulated sounds, rarely heard in Broca's dysphasics, but also not infrequent in Wernicke's dysphasia.

Varying the semantic class of target words has much more dramatic effects in conduction dysphasia than in other categories. In particular the shift from repeating nouns to repeating numbers produces a total change in the character of performance. The sound transpositions and phonemic intrusions heard with nouns usually give way to perfect production of one-word numbers. With multi-word numbers the conduction dysphasic is likely to produce verbal paraphasias—e.g. for "eight hundred and forty-seven" he may respond "eight hundred and fifty three", sparing the beginning of the number. The repetition of isolated grammatical words: "the", "of", "from", "because", "with", etc. is also more often impaired in conduction dysphasia than in the other forms.

*Sense vs nonsense:* Unique among severe dysphasics is the ability of the transcortical dysphasic to repeat faithfully polysyllabic nonsense words or words spoken in a foreign language—tasks which are the most difficult for conduction dysphasics. Nonsense words graded in length may include "flum, prannis, swikker, tarboki". Foreign words may include terms like étoile, ananas, or others of the examiner's choice.

*Multiword stimuli—length and syntactic form.* First, it is useful to test the patient with word strings of graded length—using digits and nouns, e.g.

        4—9              table—book
        8—7—1          glass—market—house
        9—6—27        garage—tape—cow—time

Here, accurate repetition span approaching the normal length of six to eight digits and five to seven nouns are obtained only from transcortical dysphasics. Span for words is usually markedly below normal levels in all other forms of dysphasia.

Sentences for repetition should be designed carefully to control quantity of lexical information, the extent to which they tap variations in syntactic structure, and whether their predominant composition consists of contentive or of grammatical function words. Prototypical examples are given here; the examiner can construct others along the same lines.

*Length of sentence.* In grading sentences for length it is well to keep to a standard subject-verb-object form in a simple past or present tense, adding adjectival or adverbial modifiers for increasing the amount of information to be retained. Thus, length of span is not confounded with the ability to process and reproduce grammatical forms.

Examples:    The boy threw the ball.

The black dog chases the gray cat.

The three women walked to the market together.

The crowded liner steamed into port under a blue sky.

*Sequential probability and vocabulary level.* Another factor to be controlled in the testing of sentence repetition is the degree to which the successive words in the sentence are predictable from the preceding words. Predictability is heightened by using words of high frequency in a semantic frame which constrains the choice of vocabulary. The Boston Diagnostic Aphasia Exam (Goodglass and Kaplan, 1972) specifically controls this factor with contrasts of the following type: At the seven-word level we have.

Near the table in the dining room (High Probability)

versus

The barn swallow captured a plump worm (Low Probability).

*Grammatical composition of stimuli for repetition.* We have found empirically that conduction dysphasics are particularly vulnerable to sentences composed predominantly of small grammatical words.

*e.g.*  No ifs ands or buts

We would have been there by now.

Sentences of this type are not disproportionately difficult for other patients and are repeated perfectly well by transcortical dysphasics.


# e) Tests of Word Retrieval

The ability to name is a fundamental aspect of language and its disturbance generally serves as a non-specific index of the degree of severity of dysphasia. In general, too, the accessibility of a word for a patient is comparable across modes of stimulation—whether on picture presentation, tactile presentation, or purely auditory (in the case of objects such as 'watch' that can be recognized by sound or in response to definition). Nevertheless there are individuals who show selective variability in word retrieval as a function of the mode of stimulation (*cf.* Spreen, Benton and Van Allen, 1966; Goodglass, Barton and Kaplan, 1968).

*Oral word reading.* For fairly obvious reasons performance on this task is most likely to deviate from other types of stimulation. If the ability to decode written language is selectively damaged, the patient is unable to recognize or produce the word, regardless of his naming ability. On the other hand, it is possible for some patients, notably transcortical dysphasics, to translate the written code into speech though they may read aloud without comprehension. One advantage of testing word retrieval through graphic input is that it is equally easy to test for any class of word, whether abstract or concrete, verb, adjective or grammatical function word. For this reason it is well to test the ability to read aloud words of all parts of speech. Some agrammatic patients immediately recognize words like "for , "but", "with", "the" as in the class of words they cannot verbalize. Differences in the accessibility of other words as a function of their semantic class (*e.g.,* number versus objects) may also be reflected in patients' ability to read these words aloud.

Various types of failure in this task reflect the wide variations in the process underlying oral reading in different patients. Some patients, unable to retrieve the word as a unit, make partially correct oral renditions of the phonic values of the letters. Thus, "yellow" may be misread as "yettow" or "nation" as "nattin". In an error of this type, it is usually the word beginning which is preserved and the end which suffers. In some instances the word is misread as another which has the same opening sounds—*e.g.,* "trip" for "train". Whereas in the foregoing instances, the patients' erroneous attempts are guided, at least in part, by phonetic decoding, some patients are prone to a totally different error process—a semantically based one. In these instances a word semantically related but acoustically unrelated to the written target may be produced. For instance, shown the word "bench" the patient may read "chair", shown "green" he may read "blue". These phenomena suggest that the written word may directly arouse a semantic associate, without the intermediary of phonetic transposition. It is a matter for conjecture whether the graphic symbol arouses the concept of its referent specifically only to have the patient misname it through a process similar to the semantic paraphasia of object naming. Another possibility is that the written word merely arouses a semantic field, which in turn triggers the production of the most randomly available word sharing that association. The latter appears to be the case in the instance of the misreading of the small grammatical words, *e.g.,* when, as often happens, the patient reads "to" for "with" or "by" for "from", he appears to be responding only to a recognition of the category of grammatical function words, with no sense of the semantic value of the particular target stimulus. The phenomenon of misreading based on semantic rather than phonetic features has been termed "phonemic dyslexia" (Warrington) or "deep dyslexia" (Marshall and Newcombe, 1973) and its implications for the normal reading process have been reviewed by a number of investigators (Patterson and Marcel, 1977; Saffran, Schwartz and Marin, 1976).

*Responsive naming.* While oral word reading can sometimes be carried out without any appreciation of the word's meaning, naming in response to a defining question involves true word retrieval in response to the arousal of its concept. The procedure in this approach is to probe for the availability of a target word by a question which refers to one or more criterial features of the target, *e.g.,* for elephant, one may ask "What is the big animal that has tusks and a long trunk?" For "razor", "What do you shave with?" For "12", "How many things make a dozen?"

It is important to bear in mind that performance on this task presupposes an adequate level of auditory comprehension. A deficit in comprehension may result in a lower rating in responsive speech than in naming to picture confrontation.

*Visual confrontation naming.* The ability to name objects or pictures on visual confrontation is perhaps the best single index of the overall severity of dysphasia. There are of course, strong exceptions to this rule of thumb. Transcortical motor dysphasics may name very well but be unable to formulate any speech in free conversation. On the other hand, some anomic patients may communicate fairly effectively with the aid of circumlocutions although they fail badly on naming tasks. These exceptional instances are relatively uncommon, however.

While it is desirable to have a set of test cards for naming, such as are provided in any published dysphasia test, an adequate array of objects for naming can almost always be improvised from among articles and furnishings which are always at hand, including parts of the body.

The examiner is not merely interested in counting the number of correct responses but the distribution of the failures, the character of the errors which occur, and the patient's ability to profit from cues offered by the examiner. With respect to the distribution of errors we note whether word-finding failures occur scattered among words of relatively high frequency, or whether the patient is fairly consistent in his success until he reaches a ceiling with relatively uncommon words. Next, how does the patient fail? Typical of the Broca's dysphasic is the recovery of the initial sound and some fragments of other parts of the word, often conveying some sense of the length of the word. Also characteristic of the Broca's dysphasic is labored and inaccurate phoneme production, which makes the output difficult to transcribe. In line with these indications that the target word is close to being retrieved, the Broca patient has a high percentage of successes when offered the initial sound of the target word as a cue. The quality of paraphasic errors in a naming test is also different among varieties of dysphasic patients. Sound transpositions and substitutions (phonemic paraphasia), are most typical of patients with conduction dysphasia. For example, shown a picture of a "zodiac" a patient of ours produced "zokiad" which he was unable to improve in the course of repeated efforts at self correction, nor when finally given a model for repetition. As opposed to the poorly formed individual sounds of Broca's dysphasics, phonemic paraphasias are freely and clearly produced and are easily transcribed. Semantic (or verbal) paraphasias (*e.g.* "rocker" for "cradle") may occasionally be produced by patients of all types, but they are most frequent in the production of Wernicke's dysphasics, who also produce phonemic paraphasias, as well as instances of totally neologistic utterances or words in which but a fragment of the target can be detected—the remainder being totally confabulated. The foregoing types of errors are rarely heard from non-dysphasic patients—even those with some reduction in word retrieval capacity as a manifestation of memory disorder. It is common for both dysphasics and non-dysphasics to give functional definitions in lieu of an object name: *e.g.* "you look far away" for a picture of a telescope; it is also common to obtain a superordinate term like "animal" in place of a specific target like "camel" or a nominalized verb like "sweeper" in place of broom. Although these forms of error may be obtained from non-dysphasics for words which are at the fringe of their working vocabulary, they may also appear as strategies of dysphasics for words which were readily accessible to them premorbidly.

It is not sufficient to test the ability to name objects alone since many patients have selective deficiencies or selective retention of such special categories as numbers, letters of the alphabet, colors, verbs, and body parts. The examination materials should therefore include stimuli from all of these classes.

Research on naming reveals that the vast majority of dysphasics show identical naming disorders whether tested by visual confrontation, tactile presentation of objects, by hearing characteristic sounds or even through the sense of smell (Goodglass, Barton and Kaplan, 1968). When tactile and visual presentation are dissociated, failure through the tactile modality is more common than the reverse (Spreen, Benton and Van Allen, 1966). Before attempting to interpret a deficit in tactile naming one must ascertain whether the patient has adequate recognition through the tactile mode. Tactile recognition may be impaired because of impairment in the primary modalities of touch and position sense—a defect which, if present, would be confined to the hand contralateral to the brain lesion. It may be due, alternatively, to a tactile agnosia—loss of tactile recognition which sometimes results from a left parietal lesion, in spite of preserved primary touch and position sense. In either case, the patient's tactile recognition can readily be tested by letting him feel an object with his eyes shut, then select it visually from an array of objects.

Once tactile recognition has been established, failure of tactile naming from the left hand alone indicates a lesion either in the corpus callosum or deep in the hemisphere in such a way as to interrupt callosal fibres. Failure of transmission of tactile information from the right hemisphere to the left prevents associating this sensory input with a verbal label. Analogous disorders of naming restricted to visual input (optic dysphasia) or to auditory input (acoustic dysphasia) are very rare. In either case it is first necessary to determine that the patient does not have a visual agnosia or an auditory agnosia, which would prevent his recognizing the stimulus—a prerequisite to being able to name it.

*Word lists.* The ability to give lists of words within an assigned category (*e.g.* "animals") is useful as an index of improvement in dysphasic patients. This task in particularly vulnerable to frontal lobe disease, as well as to extensive lesions elsewhere in the brain, so it is not highly pathognomonic of dysphasia, and its value as a sensitive measure is applicable in other cerebral disorders as well. Normal adults give an average of 18 animals in a minute. Other useful categories are makes of automobiles, articles of clothing, names of occupations.

## f) Reading

Reading is a complex achievement involving the parallel operation of several perceptual-linguistic processes, and dysphasia, in its various forms, may serve to dissociate these processes—sparing some and damaging others. In so doing it reveals the possibility of gaining at least partial meaning through the level of global associations of a whole word to its semantic referent as well as the opposite process of pronouncing any English letter combination without being able to extract meaning from it. Reading may display just as vividly as speech, the selectivity of agrammatism for substantive words, leaving functors and verbs unpronounceable and not understood.

It would be erroneous to expect to find some dissociations of the types described in every case of dysphasia; in fact most dysphasics have a relatively undifferentiated reading deficit which varies in degree of severity chiefly in terms of frequency of vocabulary, length of message, and level of inference required to gain meaning. However, the content of the examination should be designed to permit detection of any of the selective deficits. Some of the recommended procedures are described below.

*Symbol discrimination.* The first step in evaluating the status of reading in the severe dysphasic is to determine whether he recognizes the identity of various forms of the alphabetic symbols. (This of course presupposes normal premorbid ability to read print and longhand.) Some clue to this ability will already have been obtained from the patient's ability to select letters from oral presentation. However, even when a patient cannot match a letter with its spoken name, he may be able to match it to another written form. A simple test of this ability is to present a stimulus letter (or familiar short word) on a card in one style of print (block letters, or lower case, or longhand) and give him a multiple choice of five items which are visually confusable with and include the correct response, written in a different style from the stimulus. This is an easy task and patients who fail it completely are, with few exceptions, profoundly alexic. Yet there are exceptions: patients who are unable to recognize letters in isolation and who may later show comprehension of whole words.

*Phonetic matching.* In testing the ability to find the graphic representation of a spoken word or syllable, we must keep in mind the possibility of another process—responding to the meaning of the written word without regard to its phonic structure. One approach, used in the Boston Diagnostic Aphasia Exam, is to offer the patient a multiple choice array which includes choices related in connotation to the target as well as choices related visually to the target. As an example, one item requires the patient to select a match to the spoken stimulus "puppy" in an array consisting of the words bunny, puppy, kitten, playful, hippo. The patient responding directly to connotation may choose "kitten" or "playful".

A more direct test of phonetic matching ability is to present a multiple choice of written nonsense syllables, one of which corresponds to a spoken stimulus. This type of test can be made easy by making the foils very different from the target or increased in difficulty by varying only a single letter, which may be the initial or final consonant or the vowel.

*Word-concept matching.* The foregoing tests of phonetic matching can be solved without any comprehension of meaning. The simplest way of testing for written word meaning is to have the patient match written words to pictures. A single word may be shown at a time with a multiple choice of pictures, or a picture may be matched to one word of a multiple choice. When a patient fails to demonstrate the ability to make a word selection under these conditions we may infer that he is severely alexic. However as noted previously, even a severely alexic patient may derive some sense from the written word even if only at the level of connotation. In the preceding section we described a test which permitted matching the meaning derived from the spoken stimulus to a written word in the same connotative category. A variant of the earlier procedure is to present several words of widely differing affective tone: *e.g.*, "tiger", "rose", "pillow", "snow" and ask for identification with an affectively toned description—*e.g.* "a dangerous wild animal", "something soft and comfortable", etc. Failing to demonstrate any comprehension by this means, which depends on the comprehension of oral input, one may still be able to show some partial appreciation of meaning by use of the "odd word out" test (Albert *et al.*, 1973). A short (4—5 word) list of words belonging to the same category except for one extraneous one is presented and the patient is asked to find the one that is "different and doesn't belong here". For example he may be shown wolf, bear, lily, lion, tiger, from which he is to select "lily". It may be necessary to assist the patient to grasp the intent of the task by providing one or two preliminary examples, which the examiner solves, using suitable pantomime.

*Understanding oral spelling.* The ability to understand words as they are spelled by an examiner is sometimes paradoxically lost in the presence of adequate reading performance. However, it is perfectly well preserved in one class of severe alexics: those with pure alexia but retained writing. These patients have a selective dissociation of visual input from the language decoding system, but can "read" through non-visual modalities. Thus, they can understand sentences spelled aloud to them word by word and even understand letters traced on their palms. While the status of oral spelled comprehension is of some interest in all dysphasics, it is a critical diagnostic indicator when pure alexia is suspected.

The test procedure is straightforward, except in the case of patients whose impaired expression prevents them from saying what they understood the word to be. In these instances the examiner offers a series of words orally to see of the correct one is accepted. In administering the test it is well to start with primer words which are recognized by rote (*e.g.* "cat", "man", etc.) and build up to longer words.

*Understanding connected text.* The simplest procedure testing reading comprehension at the bedside is to present written commands to be carried out. However, this technique can be trusted only when the patient performs correctly or performs partially, but incorrectly. Failure to attempt a response is often encountered in patients who are found to show adequate reading comprehension when tested by other means. It would appear that these patients have great difficulty in grasping the notion that the command is not only to be read, but also to be performed. Their failure, then, lies at the level of attaining a proper mental set, rather than in the comprehension of the content of the message.

Empirically, we find that the optimum technique is to present sentences in which the final word or phrase is omitted, to be selected by pointing to a multiple choice array. An example, taken from the Boston Diagnostic Aphasia Examination is:

A MOTHER HAS A ................

COOK     CHILD     BAKING     WASH

This type of item can be elaborated to paragraph length and advanced levels of difficulty. By having all of the choices at least peripherally associated to the target item, one can be sure that the item is not solved by merely eliminating the choices which are grossly unrelated to the sense of the sentence.

## g) Examination of Writing Ability

Writing ability is parallel to speech production in that it can be broken down into the recall of motor execution (or mechanics) of writing (analogous to articulation), recall of individual written words (analogous to word finding) and sentence writing (analogous to syntax). However, writing is more complex than speech because words can be recalled as motor automatisms, as transcriptions from subvocal spelling, as transcription from sound, based on the rules of phonics, as visually guided configurations, and as the resultant of all four of the foregoing channels.

*Mechanics of writing.* As an instance of the most automatized possible writing performance the patient is asked to write his name and then his address. It is common to find that severely impaired patients are unable to sign their names in the usual longhand but are only able to produce their names in block printing. If the patient cannot sign his name in longhand on request, he should be asked to copy it from a model in longhand. A further test of the ability to recall the movements of writing is to ask the patient to transcribe a test sentence into longhand, from a model in capital letters, *e.g.*

THE QUICK BROWN FOX JUMPS OVER THE LAZY DOG.

*Recall of written symbols.* Having seen the level of writing mechanics in a purely copying context, we now ask whether the patient can recall the symbols of writing (both letters and numbers) first in an automatic series and then in response to dictation. The patient should be asked to write the alphabet and the numbers from 1 to 20. Next a random series of individual letters, and individual one and two place numbers should be given, one at a time. The patient should then be tested for his ability to write such 2 and 3 letter primer words as "the", "of", "boy", "man", "cat", etc.—words which are so overlearned that they are probably written essentially by rote, with minimal reliance on the mapping of sound sequence onto letter sequence.

*Written word recall.* A sampling of the patient's ability to retrieve the written form of words of graded difficulty should be undertaken. The first procedure is simply spelling from dictation. It is useful to select words from some standardized spelling test such as the Wide Range Achievement Test (Jastak and Jastak, 1976). Writing is a complex operation in that many avenues of memory converge to produce the final performance. Patients who were well educated and accustomed to write a great deal have a considerable repertory of words which they can write through recall of the motor sequence—just like letters and primer words. This channel is aided by visual recall of the configuration of the word. In parallel with these processes is the reconstruction of words by the conversion of the sound pattern to the corresponding letter sequence. Finally, some patients may be noted to spell words aloud and copy the letter sequence from their own oral production.

In order to ascertain to what extent patients can rely on their phonic ability—*i.e.* the ability to convert sounds to graphic sequences, it is useful to dictate some one and two syllable nonsense words, *e.g.* CASS, BOME, FLISK, CRAFTER, etc. In order to assess the patient's recall of words as a total configuration, the spelling list should include such non-phonetic items as "sugar", "weight", "cough", etc.

The second approach to tapping written word recall is by presenting object-pictures, colors, and numbers to be written. In fact, here for the first time we examine the ability of the patient to express meaning in written form, since the prior test (writing to dictation) can conceivably be done without any awareness of word meaning.

*Writing connected text.* The patient's ability to write connected text should be examined both by dictation and by free narrative writing. It is well to begin with non-dictated narrative, through the use of a standard situation-picture which the patient should describe. Alternatively he may be asked to write a narrative concerning his illness and activities in the hospital—as though writing a letter to a friend. Dictated material may consist of a series of sentences of increasing difficulty or of a standardized paragraph.

It should be noted whether the patient can use normal grammatical structure or whether he reduces his sentences to a string of isolated nouns and verbs, analogous to agrammatic speech. Many patients with Wernicke's dysphasia produce a fairly fluently written jargon, which is quite parallel in structure to their speech output. In some instances the pattern of writing may be quite disparate from that of speech—the written production of a fluent dysphasic resembling that of an agrammatic patient. In recovered dysphasics one detects little more than occasional misspelling and occasional omission of inflections or of small grammatical words.

It may be helpful to use the five point rating scale, extracted from the Boston Diagnostic Aphasia Exam to characterize the patient's narrative writing.

### Scale for Narrative Writing

No relevant writing
Isolated words or small groupings
Incomplete but relevant sentences
Unduly simplified but correct sentences
Full description

# 5. Testing for Dyspraxia

Disorders involving the execution of purposeful movement—particularly pretended representational actions—are commonly seen with dysphasia and occasionally seen in the absence of dysphasia. The absence of a strong correlation between the severity of dyspraxia and severity of dysphasia suggests that there is no causative relationship between the two but that both language and praxis are independently vulnerable to injury in the left perisylvian zone. Dyspraxia may be so severe as to prevent a patient from pointing to objects or from carrying out verbal commands; in these instances the testing of auditory comprehension is restricted to the use of yes-no questions.

Since purposeful movements involving the oral-respiratory apparatus (bucco-facial praxis) and the use of the hands and limbs may be independently affected, these two factors should be separately controlled in the examination. Further, movements in which limb activity is subordinated to bilateral movements of the body axis (axial movements) should be treated as a separate category—as discussed earlier in the section on auditory comprehension of commands.

An outstanding feature of dyspraxic disorders is that their severity is closely tied to the degree of contextual support from external stimuli. The most severe impairment is under conditions of pretended action to verbal command. The opportunity to imitate the examiner's performance represents an intermediate level of difficulty and often results in some improvement

| | | Mode of Testing | | |
|---|---|---|---|---|
| *Type of Movement* | *Specific Examples* | Oral Command | Imitation | Real Object |
| Buccofacial | Blow out match | Yes | Yes | Yes |
| | Lick lips | Yes | Yes | — |
| | Cough | Yes | Yes | — |
| | Sniff a flower | Yes | Yes | — |
| | Sip through straw | Yes | Yes | Yes |
| Conventional intransitive limb gestures | Wave goodbye | Yes | Yes | |
| | Beckon to "come here" | Yes | Yes | |
| | Salute like a soldier | Yes | Yes | |
| Use of objects | Comb hair | Yes | Yes | Yes |
| | Brush teeth | Yes | Yes | Yes |
| | Hammer a nail | Yes | Yes | Yes |
| | Saw a board | Yes | Yes | — |
| | Flip a coin | Yes | Yes | Yes |
| | Use screwdriver | Yes | Yes | Yes |
| Axial commands | Treated under auditory comprehension of commands. | | | |
| Non representational hand positions (use left hand, all fingers extended and held together). Imitation only. | a) Hand upright under chin, with thumb against neck, palm facing right | | | |
| | b) Thumb against right cheek, palm facing back | | | |
| | c) Little finger against left cheek, palm facing back | | | |

in performance. The opportunity to manipulate or respond to a real life object or situation may cause the difficulty to disappear. In other words, the order of difficulty of praxis testing from hardest to easiest is (1) pretended action to verbal command, (2) imitation, (3) manipulation of actual object. Needless to say, if poor performance can be accounted for by paralysis, weakness, or incoordination, a diagnosis of dyspraxia cannot be made. Further, if impaired auditory comprehension has been detected, it becomes difficult to interpret failure to carry out actions to oral commands and more weight must be assigned to performance under imitation.

In the course of testing for dyspraxia, it is useful to examine the ability to imitate non-representational limb movements, typified by various placements of the hand to the face, as described on p. 49.

## 6. Beyond the Formal Examination

It is well to turn once more to the notion of the dysphasia examination as a free ranging probe of the patient's capacity to function with language. In the course of reviewing methods of undertaking this exploration, one should resist the temptation to adopt the suggestions offered here as a series of formulas to be followed, as in a cookbook. The fact is that the possibilities for new and revealing procedures are practically unlimited and with their invention can come discoveries which uniquely characterize the language of certain patients or, even more importantly, which can be demonstrated repeatedly in other patients.

# Part II
# Clinical Features of Dysphasic Syndromes

# Introduction  A

This section describes the signs and symptoms of language disorders that are encountered most frequently in the clinical setting. In general, it follows the outline of dysphasia syndromes offered years ago by Wernicke (1874, 1885—1886) and Lichtheim (1885), and recently reformulated by Geschwind, Goodglass, Benson, and their associates (Geschwind, 1965; Goodglass and Kaplan, 1972; Benson and Geschwind, 1976; Benson, 1979). This system of classification has been chosen largely because it is by far the most commonly used method of grouping dysphasic patients to be found in the recent literature. There have been other published and widely used classifications of dysphasia, notably those of Head (1928), Weisenburg and McBride (1935), and Luria (1970), and where they correspond to the syndromes described in this chapter, the similarities will be pointed out.

We emphasize from the outset that the syndromes discussed in this section, regardless of whether for historical reasons they carry the name of a famous neurologist or an anatomical or pathophysiological label, are to be considered as referring to specific patterns of altered speech and language behavior, not to anatomical location or pathophysiological process. For example, fluent and nonfluent dysphasia, though highly correlated with lesions of the posterior and anterior speech zones, respectively, are not synonyms for posterior and anterior dysphasia. When speaking of Broca's and Wernicke's dysphasia, we describe a constellation of speech and language symptoms rather than implying a specific lesion location. Similarly, although the anatomical and pathophysiological terms "transcortical" and "conduction" are of historical importance and may reflect the pathophysiological mechanisms underlying some instances of these forms of dysphasia, the supporting evidence remains inconclusive. It also should be kept in mind that Broca's and Wernicke's areas are anatomical conventions with no clearly agreed upon boundaries (Bogen, 1976). The mixing of anatomical and behavioral terms may serve at times as a useful mental shorthand, but the indiscriminate mingling of anatomical, physiological, and psycholinguistic levels of description may lead to erroneous clinical conclusions in individual patients, faulty reasoning in dysphasia research, and confusion on the part of students and others who are not familiar with the terminology. Kinnier Wilson (1926) stated the problem clearly in 1926:

"The function of speech is an intellectual function; the neural arrangements underlying its activity constitute a physiological mechanism; and the component units of the latter have an anatomical localization or site. Thus, the problem can be and often has been approached from anatomical, physiological, and psychological sides, respectively and separately, and the investigators have not always been careful to avoid inappropriate transfers of terms or even of conceptions from one system to another."

If it is kept in mind that the syndromes described in this section refer to symptom complexes, this type of confusion will be avoided.

In recent years the technique of computerized axial tomography (CT scanning) has enabled the clinician to visualize the location, size, and etiology of dysphasia-producing lesions in the living patient. It is no longer necessary for the clinician to outlive his patient in order to relate anatomical and patho-logical variables to specific speech and language deficits and modes of recovery. Systematic studies of CT scan correlates of dysphasia have already substan-tiated some long-held notions of cerebral localization but cast doubt on others (Naeser and Hayward, 1978; Kertesz, 1979). The anatomical correlates of the various syndromes of dysphasia are in a process of redefinition. The use of CT scans has already changed our concepts of the patho-anatomical basis of Broca's dysphasia (Mohr, 1975; Mohr et al., 1977) and has been instrumen-tal in relating new dysphasia syndromes to underlying brain damage, such as that seen with left thalamic hemorrhage (Mohr and Duncan, 1975; Rubens, 1977). The even newer technique of positron emission tomography identifies functionally active brain regions that have been differentially tagged with radioisotopically labeled compounds of glucose that are incorporated into meta-bolically active parts of the brain during specific language, cognitive, and sensory-motor tasks (Reivich et al., 1979). It may soon be possible, with this technique, to produce a brain map relating language function to anatomical location, and to identify regions of the brain most responsible for restitution of function after brain damage. There is no question, therefore, that our concepts of the mechanisms involved in the production of and recovery from dysphasia will be significantly modified in the near future.

At the present time it remains useful to refer to the classical dysphasia syndromes. The terminology has been employed in an increasing number of papers in the recent literature. The syndromes carry with them a certain degree of anatomical predictability which, though far from perfect, allows clinicians to infer location, etiology, and prognosis. The syndromes, along with their implied causative mechanisms, may be looked upon as working hypotheses from which predictions can be made that are then amenable to verification. It is not necessary or advisable to attempt to force every patient into one of the classical syndromes. In fact, it is our own experience and that of Prins, Snow, and Wagenaar (1978) that no more than twenty or thirty percent of dysphasic patients will fit neatly into one of the specific dysphasia syndromes. It is productive, however, to recognize the individual elements that make up the major syndromes because each of these and their various combinations carry a significant degree of localizing reliability and theoretical importance.

In consideration of specific dysphasic syndromes the time post-onset and the etiology of the disturbance must be taken into account. Acute focal lesions are often associated with the phenomenon of diaschisis which lasts days to weeks and which obscures the focal characteristics of the deficit by producing widespread cerebral dysfunction and variability of performance. Syndromes seen in relatively pure form in the subacute stage of illness may later lose their distinctive features and evolve towards a nonspecific anomic dysphasia in the more chronic period (Kertesz and McCabe, 1977). Global dysphasia evolves into Broca's dysphasia in many instances (Mohr *et al.*, 1978). When conduction dysphasia and Broca's dysphasia are seen as the initial feature of a self-limited pathological process such as stroke, the potential for rapid recovery is great. Transcortical motor dysphasia often presents a dramatic speech and language disturbance in the acute period but commonly resolves over several months (Rubens, 1975). Brain tumors grow slowly and spread among normal nerve cells without actually destroying them. They are not accompanied by the more widespread cerebral dysfunction produced by diaschisis, and it is not uncommon for large tumors located within the speech zone to be associated with little in the way of dysphasic symptoms, except for a nonspecific naming deficit. Smaller circumscribed cerebral hemorrhages tend to track along white matter pathways and to distort more than destroy brain tissue, and subtotal and rapidly improving syndromes are, therefore, more common.

Finally, the effects of age and inherited and acquired premorbid individual differences in cerebral organization, particularly lateralization of function, interact with lesion variables to produce in each patient a unique combination of language, emotional, and cognitive assets and liabilities, regardless of into which specific syndrome subtype they fall. Yet, while it is true that no two dysphasic patients are exactly alike, there are clinically observable patterns of language dissolution associated with focal damage to specific parts of the language-dominant hemisphere. It is the purpose of this section to describe and characterize these constellations of symptoms.

The following table outlines dysphasia syndromes, grouping them according to whether repetition is disturbed or spared, and separating syndromes in which impairment is limited to reading and/or writing performance. Impaired repetition is strongly correlated with perisylvian pathology. Two syndromes, pure word deafness and aphemia, are not truly dysphasic syndromes because language *per se* is not impaired; but they are included among the first group of disorders because repetition is severely impaired and because the causative lesion is located in the dominant left perisylvian region. Distinguishing features of the major forms of dysphasia are outlined in the second part of the table. Various combinations of impaired performance on repetition and comprehension tasks and the degree of fluency of spontaneous speech are sufficient in themselves to distinguish the major dysphasic syndromes.

Table 1. *Clinical syndromes of dysphasia*

Dysphasia with repetition disturbance (perisylvian)

    Broca's dysphasia
    Wernicke's dysphasia
    Conduction dysphasia
    Global dysphasia

    Pure word deafness         Not truly dysphasic syndromes, but included
    Aphemia                because of defective repetition and perisylvian
                         location of lesion

Dysphasia without repetition disturbance

    Transcortical motor dysphasia (anterior isolation syndrome, dynamic dysphasia)
    Transcortical sensory dysphasia (posterior isolation syndrome)
    Mixed transcortical dysphasia (isolation of the speech area)
    Anomic dysphasia
    Dysphasia with left subcortical hemorrhage

Disturbances of reading and/or writing

    Dyslexia with dysgraphia
    Dyslexia without dysgraphia (pure word blindness)

Table 2. *Distinguishing features of major forms of dysphasia*

| | Fluency | Repetition | Comprehension | Naming |
|---|---|---|---|---|
| Global Dysphasia | Nonfluent | − | − | − |
| Broca's Dysphasia | Nonfluent | − | + | − |
| Transcortical Motor Dysphasia | Nonfluent | + | + | − |
| Mixed Transcortical Dysphasia | Nonfluent | + | − | − |
| Wernicke's Dysphasia | Fluent | − | − | − |
| Transcortical Sensory Dysphasia | Fluent | + | − | − |
| Conduction Dysphasia | Fluent | − | + | − |
| Anomic Dysphasia | Fluent | + | + | − |

− Impaired
+ Normal or relatively spared

# Neuroanatomical and Neurophysiological Considerations **B**

It is traditional that a review of the anatomy of language disorders commences with a summary of the history of the disputes between the proponents of the two main streams of thought about brain-language relationships. The ebb and flow of popularity of various localizationist and globalist theories is colorful and instructive, but it is well reviewed elsewhere (Hecaen and Albert, 1978). This issue is no longer relevant in the sense considered by previous generations. Clearly brain functions are not equally represented in all regions. Interhemispheric differences exist in 1) the perception and manipulation of higher level sensory information, 2) the organization of axial, limb and buccofacial movements and 3) the ability to generate speech and language. These interhemispheric differences are based in part on anatomical asymmetries which are evident in fetal life (Wada et al., 1975). There is intrahemispheric specialization in brain function as well, and within the left hemisphere, much of this specialization in function constitutes the anatomy of language. Even many ostensible critics of the localizationist theories resorted to a system of language classification that carried implicit functional localization which strongly resembled the classical formulations of Wernicke (1874) and Déjerine (1914). For example, Marie (1917), Head (1926), and Goldstein (1948), despite their reputations as antilocalizationists, utilized systems of classification based on functional anatomy.

The foregoing does not mean that extreme localizationist views are uncritically accepted. Some authors have proposed strict localization of remarkably specific functions into sharply defined areas of brain (Kleist, 1939; Nielsen, 1946). This extraordinary compartmentalization of supposedly independent functions has not been confirmed, nor has it been particularly useful in establishing a basis for advancing knowledge of brain-behavior relationships.

At the present level of knowledge about anatomical-functional correlations it may be more productive and appropriately cautious to consider the brain as being organized into various zones. These zones are not precisely demarcated from adjacent zones; rather they represent the focus of a brain function which may be variously influenced by neighboring regions. Many researchers have utilized such a framework for language localization, but Luria's conception of brain regions as being overlapping zones of relative specialization may best exemplify the cautious assumptions of modern functional neuroanatomy (1966).

In the broadest descriptive way, this view of the functional anatomy of language recognizes several different zones within the left hemisphere. There are regions whose primary connections are with the thalamus or with the brain stem and spinal cord. On the afferent limb, the calcarine cortex in the occipital lobes and Heschl's gyrus in the transverse temporal gyri are the primary cortical regions for visual and auditory reception respectively. Their anatomical limits are fairly well defined (Truex, Carpenter, 1969). The next level of processing involves sensory association cortex. In the left hemisphere the auditory association cortex includes the posterior planum temporale and adjacent regions in the posterior superior temporal gyrus. This region is Wernicke's area (Fig. 3, area W) and in most adults it is larger in the left hemisphere (Geschwind, Levitsky, 1968). The visual association cortex of the occipital lobes blends into the parieto-occipito-temporal junction. The association cortex is defined by its projections from primary sensory cortex but also its transcallosal connections with homologous contralateral brain (Geschwind, 1965).

Visual association cortex operates by a method of specific stepwise feature discrimination (Hubel, Wiesel, 1979) and at higher levels of discrimination by modality specific regions such as those which respond to color (Zeki, 1973). Progression into these sensory association systems is marked by a decrease in the amount of direct contact with the periphery and an increase in the amount of contact with other related cortical regions. Ultimately, there is an area of association cortex which is no longer specific for a single sensory modality, e.g. the angular gyrus (Fig. 3, area A). This gyrus stretches around the posterior limit of the superior temporal sulcus and is considered part of the inferior parietal lobule. It is the crossroads of the assocation centers of the auditory, visual and tactile systems. Intrahemispheric connections to and from the angular gyrus are largely limited to these adjacent association regions. Geschwind (1965) has detailed the evolution, the anatomical connections and the role in crossmodal associations of the angular gyrus. The anatomy of these posterior (parietal, temporal and occipital) sensory systems has been correlated with syndromes of language dissolution.

A similar hierarchy of language related brain systems exists in the left frontal lobe. The lower third of the prerolandic gyrus has direct monosynaptic connections with brainstem motor neurons. This corticobulbar pathway is the route through which rapid integration of the buccofaciolingual movements necessary for speech is controlled. Extensive intrahemispheric connections to this region from motor association cortex suggest that the association cortex is in position to modulate and integrate multiple step, multiple muscle movements such as those required for speech. The region in the inferior prerolandic region of the frontal operculum of the left hemisphere is called Broca's

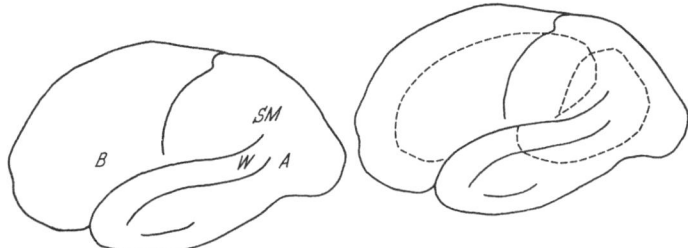

Fig. 3. Stylized depiction of lateral convexity of the left hemisphere. *B* indicates Broca's area immediately anterior to inferior motor strip. *W* indicates Wernicke's area in posterior superior temporal gyrus. *A* indicates the angular gyrus at the temporoparietal junction. *SM* indicates the supramarginal gyrus in the inferior parietal lobule. The figure on the right represents the approximate areas of involvement in nonfluent dysphasias with retained comprehension (left dashed circle) and in fluent dysphasias with or without retained comprehension (right dashed circle); lasting global dysphasias would occur with a lesion encampassing both circles

area (Fig. 3, area B). As with sensory association regions this motor association cortex is extensively connected to homologous regions of the right hemisphere by transcallosal pathways. Posterior (sensory) association areas have important projections to frontal association cortex (Geschwind, 1965). The frontal lobe anterior to the operculum does not have the same cellular anatomy as Broca's area and the precentral gyrus, and motor speech functions are not as obvious. There is evidence, however, that extensive portions of this premotor cortex are involved in language production (Milner, 1964). The anatomy of the anterior left hemispheric motor and premotor systems correlates with certain syndromes of dysphasia.

This simplified summary of cortical regions, their intrahemispheric and interhemispheric connections and the relevance of these structures to language disorders borrows from three sources. The clinico-anatomical studies of Déjerine (1914) based on reported and personally examined cases established the foundations for localization of speech and language functions. The "zone of language" had three primary regions: 1) an anterior section, Broca's area; 2) a temporal section, Wernicke's area; 3) a posterior portion, the angular gyrus. These were thought to represent centers for motor speech, verbal auditory images and verbal visual images, respectively. Geschwind (1965) reviewed the wealth of animal and human data on intracerebral connections and described the anatomical bases of single modality language disturbances (pure dyslexia, pure word deafness and pure word dumbness). In addition, the concept of an association area, the angular gyrus, which is determined by its interconnections between all sensory modalities, suggested a ready explanation for dyslexia with dysgraphia as a disorder of crossmodal connections. Finally, identification of specific intrahemispheric connections between posterior and anterior language association areas (that is, between Wernicke's area and Broca's area), revealed one mechanism for the syndrome of conduction dysphasia. Luria (1966) provided an anatomico-linguistic classification of dysphasia which most directly modifies the ideal of cortical centers. Recognizing that

as behaviors become increasingly complex, precise localization of brain function and exact demarcation of brain centers decrease, Luria outlined a hier-
archical system of functional organization. Large portions of the left frontal
lobe are involved in transformation of thought into sensible and sequential
utterances and in the maintenance of readiness to speak. More clearly localized
frontal opercular regions are specifically engaged in the learned organization
of the motor systems of language. Finally, the well-demarcated inferior
prerolandic gyrus is the origin of the specific motor pathway which controls
the movements of speech. There is a reversed arrangement of verbal sensory
systems. They proceed from specific, well localized cortical centers of primary
sensation to less specific, broadly localized centers of crossmodal activities.
The systems of analysis and classification proposed by these three important
figures combine in a most coherent picture of brain anatomy and its relationship to language function.

Certain caveats about brain-behavior localization must be reemphasized.
Different diseases in the same area of brain may produce considerably different
clinical syndromes depending upon individual differences or upon the destruc-
tiveness of the disease, upon the course of onset and resolution, and upon the
presence of indirect causes of brain dysfunction, such as cerebral edema.
Location of the lesion may be the most important factor in production of
clinical syndromes, but attention must be paid to etiology in comparing the
effects of brain lesions in different locations. Whatever the etiology, the
chronicity of the brain lesion and the degree of spontaneous recovery or of
therapeutic successes must be considered. There is a natural progression of
change in many dysphasic syndromes, and many clinical syndromes may represent some stage in the evolution of the language dysfunction of a single lesion
(Kertesz, McCabe, 1977). Age of the patient at onset of the language disorder
also affects the clinical appearance (Brown, Hecaen, 1976). Young children
are more likely to have a nonfluent dysphasia with preserved comprehension
regardless of lesion site. Eldery patients are more likely to have a fluent
dysphasia (Obler et al., 1979). Left handedness also modifies the specificity of
brain-behavioral correlations outlined above (Goodglass, Quadfasel, 1954;
Brown, Hecaen, 1976).

These cautionary remarks are important in any consideration of the
functional anatomy of specific clinical syndromes. The language syndromes
themselves are discussed subsequently, but we will consider their anatomical
basis here.

*Broca's dysphasia* is a complex of neurological and linguistic abnormalities
produced by extensive lesions of the left frontoparietal regions (Fig. 4 A) (Mohr
*et al.,* 1978; Naeser, Hayward, 1978). Any presumption that a lesion limited to
Broca's area would result in Broca's dysphasia would depend upon knowledge
of the exact extent of Broca's area and upon agreement about the language
parameters of Broca's dysphasia. Broca's original patient had extensive lesions
of both the second and third frontal gyri (as well as parietal and temporal
regions), but Broca believed the posterior portions of the third inferior frontal
gyrus to be most important in speech. Subsequently, the critical size and
boundaries of Broca's area have waxed and waned dramatically in the litera-

ture. The polar positions are those of Marie that no part of the frontal convolutions are important in normal speech and of von Monakow (quoted in Goldstein, 1948) that the entire third frontal convolution, the entire frontal operculum and the anterior insula make up Broca's area. Mohr *et al.* (1978) have recently provided an extensive analysis of the controversy about Broca's area and Broca's dysphasia. Lesions limited to the frontal operculum or its outflow produce a syndrome of speech disturbance which is 1) nonfluent initially, 2) associated with transient buccofacial apraxia and 3) rapidly improves. A mild writing disturbance may be present. Most cases have resulted from small embolic infarctions (Mohr *et al.*, 1978) or small surface tumors (Hecaen, Consoli, 1973). Small lesions of the inferior prerolandic motor strip are responsible for a similar disorder but without agraphia (Lecours, Lhermitte, 1976). This disorder has been variously labelled pure anarthria, pure word dumbness or aphemia. The language disorder of Broca's dysphasia has usually evolved over a period of weeks or months out of a more severe nonfluent (even global) dysphasia. The lesion involves large portions of frontal lobe and parietal lobe, generally conforming to an infarction of the superior division of the middle cerebral artery (Naeser, Hayward, 1978). Most cases have resulted from large embolic infarctions (Mohr *et al.*, 1978).

*Transcortical motor dysphasia* has been reported with 3 separate pathological substrates. Most cases have large left frontal lesions which spare Broca's area (Rubens, 1976). The majority of these patients have strokes in the territory of the most anterior divisions of the middle cerebral artery (Fig. 4 B). The exact vascular pathology of these cases is not known, but the course and distribution suggest that infarction in the borderzone between anterior and middle cerebral arteries might be responsible. Large left frontal lobectomies produce a disturbance of spontaneous speech and verbal fluency which resem-

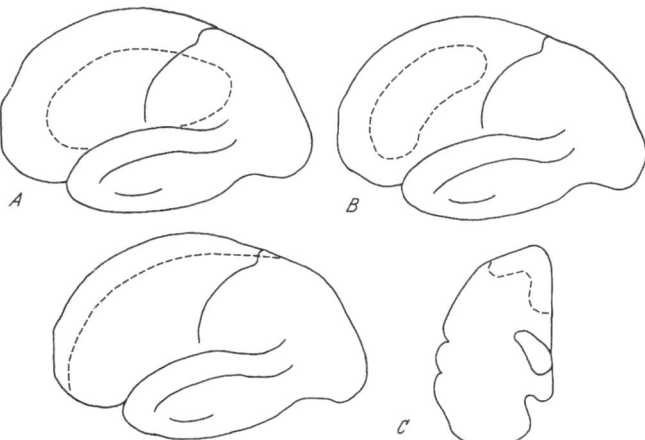

Fig. 4. *A* Typical distribution of lesion with lasting Broca's dysphasia. *B* Typical lesion of transcortical motor dysphasia; area is in anterior distribution of middle cerebral artery or at borderzone of anterior and middle cerebral arteries. *C* Distribution of left anterior cerebral artery territory infarction which may also produce a transcortical motor dysphasia

bles that in transcortical motor dysphasia. This confirms the importance of this region of the brain in facilitating and organizing language production (Milner, 1964), although the exact dimensions of the portion of left frontal lobe involved in these functions cannot be stated. Some patients with transcortical motor dysphasia have infarctions in the distribution of the left anterior cerebral artery (Fig. 4 C) (Rubens, 1975; Alexander, Schmitt, 1979). At least a portion of the speech deficit in these cases is secondary to the injury to the supplementary motor area on the medial surface of the frontal lobe. This region has important connections with the limbic system and may be involved in the initiation of speech and in the maintenance of readiness to speak (Botez, Barbeau, 1971). Diseases, other than stroke, which damage the supplementary motor area may produce a similar disturbance in speech initiation and in maintenance of normal fluency. Finally, transcortical motor dysphasia rarely may be seen with damage to Broca's area directly. The mechanism of this unusual event is not known.

The functional anatomy of the *anterior dysphasias* may be briefly summarized. Large lesions of the frontal lobe that spare the frontal operculum will result in transcortical motor dysphasia. Large lesions of the frontal operculum and adjacent frontal lobe and parietal operculum will result in Broca's dysphasia. Smaller lesions of the frontal operculum will produce a less marked disturbance in language and speech production. An even smaller lesion of just the inferior prerolandic motor strip may produce a limited disorder of articulation with normal language (aphemia or pure word dumbness) (Fig. 3, area B).

The syndrome of *Wernicke's dysphasia* is produced by lesions of the dominant posterior superior temporal gyrus. Recent anatomical studies (Naeser, Hayward, 1978; Kertesz, 1978) of patients with Wernicke's dysphasia reveal a large area of injury in the posterior temporal-inferior parietal regions (Fig. 5 A). Heschl's gyrus (primary auditory cortex) is anteromedial to Wernicke's area, and the angular gyrus is posterior to Wernicke's area. This anatomical arrangement accounts for occasional dissociations in auditory and written comprehension in some patients with a Wernicke's area lesion. This discrepancy has been dubbed a "modality bias" and reflects the fact that anteriorly placed lesions will disrupt auditory input to Wernicke's area more severly than visual (Hecaen, 1969; Hier, Mohr, 1977). There will be a more prominent word deaf quality to the language disorder with the more anteriorly located lesion. If the lesion is small enough, it may damage the ipsilateral auditory radiations to Heschl's gyrus and extend under Wernicke's area to disconnect transcallosal pathways between auditory association cortices. This small single lesion is the basis of the syndrome of pure word deafness with a single dominant hemisphere lesion (Geschwind, 1965). In fact, some cases of pure word deafness are seen in the recovered stage of Wernicke's dysphasia. This probably represents resolution of edema in temporal cortex or partial injury of temporal cortex. The focus of the lesion is subcortical and somewhat anterior to Wernicke's area. Even extensive, but more anteriorly placed, temporal lobe lesions may produce a Wernicke's dysphasia which resolves into pure word deafness. Certain etiologies of temporal lobe injury (*i.e.* herpes simplex

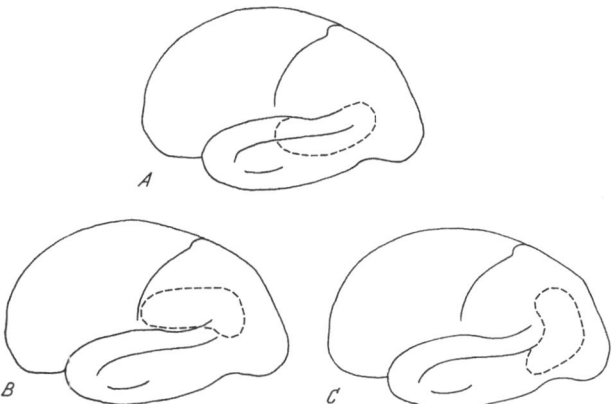

Fig. 5. *A* Typical distribution of lesion in Wernicke's dysphasia. *B* Total area in which lesion resulting in conduction dysphasia may occur. *C* Posterior temporoparieto-occipital border-zone; lesions here may result in transcortical sensory dysphasia or anomic dysphasia or alexia with agraphia depending upon size and exact distribution

encephalitis) are more likely to produce that syndrome (Hier, Mohr, 1977). Most cases of pure word deafness have bilateral lesions, but the functional disconnection is the same as in unilateral cases (Geschwind, 1965). The cortico-subcortical left temporal lesion either damages Heschl's gyrus or destroys its posterior projections to Wernicke's area. The right temporal lesion is usually subcortical and extensive enough to damage the transcallosal connections from right hemisphere to left. If the bilateral lesions are actually in Heschl's gyri and undercut the cortex, cortical deafness would result. Depending upon small differences in lesion location, these syndromes may blend. In particular, cortical deafness and pure word deafness are closely related disturbances (Earnest *et al.*, 1977).

A lesion limited to the angular gyrus (Fig. 6), posterior to Wernicke's area, will produce a disturbance almost exclusively involving written language: *dyslexia with dysgraphia*. The importance of the angular gyrus in the crossmodal associations necessary for reading was discussed above. The most common cause of this disorder is embolic infarction in the distribution of the angular artery, a terminal branch of the middle cerebral artery. More extensive infarction would involve posterior elements of Wernicke's area. This would result in the other pole of the "modality bias" for Wernicke's area: a Wernicke's dysphasia with relatively greater impairment of reading.

An even more posterior and subcortical lesion in the angular gyrus region might spare the angular cortex but damage afferent pathways from the visual association areas of both hemispheres and produce *pure word blindness (dyslexia without dysgraphia)*. Greenblatt (1976) designated this form of dyslexia: subangular dyslexia (Fig. 6). This unilateral lesion is the exact parallel of the single left hemisphere lesion producing pure word deafness: simultaneous disconnection of intrahemispheric pathways into association cortex and disconnection of interhemispheric (transcallosal) pathways to dominant angular gyrus.

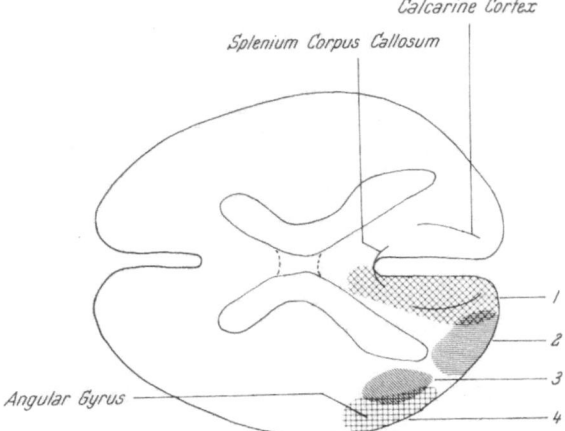

Fig. 6. Schematic depiction of horizontal section of brain with lesions which result in alexia. *1* Large medial temporo-occipital lesions with involvement of splenium of corpus callosum results in lasting alexia without dysgraphia. *2* Smaller ventrolateral occipital lesion results in transient pure alexia. *3* Subangular lesion also results in alexia without dysgraphia. *4* Angular gyrus lesion produces alexia with dysgraphia

Most of the small number of cases of subangular alexia have been secondary to neoplasm or surgical manipulation and have been transient. The classical pathological anatomy of dyslexia without dysgraphia is a left posterior cerebral artery distribution infarction (Fig. 6) (Déjerine, 1892; Geschwind, 1965). There is injury to the left visual cortex and to the splenium of the corpus callosum. The effect of this disconnection is the same as in the subangular cases: visual-verbal information cannot get into the left angular gyrus. The right hemianopia and the ventral splenium lesion, which blocks transcallosal connections from the right occipital region, combine to produce this visual-verbal disconnection.

Extensive parietotemporal lesions which spare Wernicke's area may result in *transcortical sensory dysphasia* (Fig. 5C). Projection systems to the temporal lobe being spared, audiological decoding of speech is preserved, and repetition is possible. Comprehension of the meaning of the words is defective. This syndrome is unusual and rarely follows a single ischemic stroke. An acute Wernicke's dysphasia rarely evolves into a transcortical sensory dysphasia. An infarction of the borderzone between anterior, middle and posterior cerebral arteries may produce this disorder. With an extensive posterior lesion severe dyslexia is common. Smaller lesions, limited to the inferior temporo-occipital junction, may result in a pure *anomic dysphasia* (Benson, 1979). Anomia has no localizing value by itself. Many forms of dysphasia may improve to such an extent that only anomia persists (Kertesz, McCabe, 1977) but, when pure anomia occurs in isolation as an acute language deficit, the lesion is frequently in the temporal-occipital junction or in the second temporal gyrus. This disorder is commonly the result of an embolic infarction in the terminal portions of the inferior division of the middle cerebral artery. This region may be injured at the margin of a left posterior cerebral artery territory infarction as well (Benson *et al.*, 1974).

The functional anatomy of the *posterior dysphasias* may be briefly summarized. Large lesions of the parieto-temporo-occipital junction disrupt lexical elements of comprehension and crossmodal associations (reading and writing); a transcortical sensory dysphasia is seen. Smaller lesions of this region may produce relatively isolated disturbances in written language, when the angular gyrus is involved, or in word retrieval when the temporo-occipital junction is involved. When the lesion involves the posterior portions of the superior temporal gyrus, Wernicke's dysphasia occurs. Depending upon the anterior or posterior extent of this lesion, a relatively greater disturbance in auditory or reading comprehension can be detected (Hecaen *et al.*, 1968; Hier, Mohr, 1977). With more anteriorly placed lesions of the superior temporal gyrus, particularly those with subcortical extension, pure word deafness results (Geschwind, 1965).

*Conduction dysphasia* is a fluent dysphasia with preserved auditory comprehension. The functional basis of this syndrome in some cases is believed to be a disconnection of the posterior, superior temporal language zones from anterior language zones. The major white matter tract from Wernicke's area to the frontal operculum is the arcuate fasciculus (Geschwind, 1965), and lesions of this pathway or its connections in the inferior parietal lobule (supra-marginal gyrus) may result in conduction dysphasia (Fig. 5 B). Pathologically verified cases have had lesions in areas around the posterior sylvian region from Wernicke's area to the postrolandic gyrus (Benson *et al.*, 1973); and it is possible that a common denominator of these syndromes is a lesion in the supramarginal gyrus and/or its subjacent white matter.

The importance to language of several subcortical white matter pathways was discussed above. Damage to specific white matter tracts is responsible in part for the syndromes of pure word deafness and pure alexia. Likewise, in some cases, conduction dysphasia is probably due to injury to a specific white matter pathway. The possible role in language of subcortical nuclear structures has not yet been mentioned, and is not traditionally considered.

Diseases of the basal ganglia produce speech disturbances (Darley *et al.*, 1975). These speech deficits commonly include impaired articulation, diminished volume, abnormal prosody and poor respiratory control. Whether diseases of subcortical nuclear structures produce specific syndromes of language disorder is not yet clarified. The available experience with tumors (Smythe and Stern, 1938) and spontaneous intracerebral hemorrhages (Alexander, Lo Verme, 1980) supports the contention that thalamic lesions and putaminal lesions may produce at least anomia and dysgraphia (Fig. 7). Little specificity about anatomy is available from either of these diseases. Surgical lesions of the ventrolateral nucleus and pulvinar of the thalamus suggest that these nuclei may be involved in language function, in particular in word retrieval and in short term memory. With *all* types of thalamic lesions only left sided lesions result in language disturbance. The posterior thalamic nuclei have important projections to the posterior temporal and parietal cortex; this anatomical relationship implies that the thalamic structures are involved in alerting

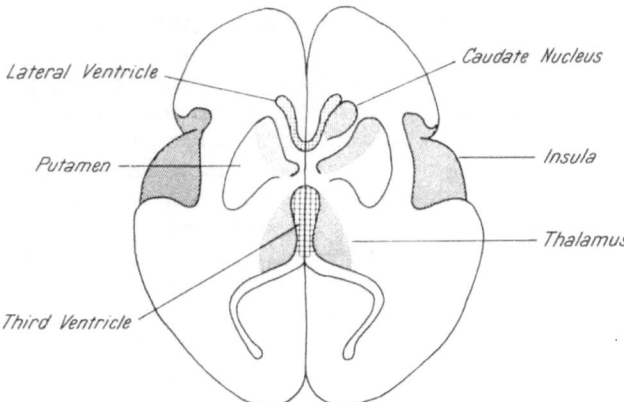

Fig. 7. Schematic depiction of horizontal section of brain identifying subcortical structures involved in language dysfunction

cortical regions for the specific tasks of word retrieval (Brown, 1977). Further investigation of the role of subcortical nuclear structures in language is a necessity.

The anatomical basis of language outlined in this chapter is founded on inferences about function following specific, well-localized brain lesions. The most useful lesions for brain-behavior correlations are single, stable infarctions. The formulations of Déjerine (1914) and Geschwind (1965) are based almost exclusively on the study of patients with brain infarction. The study of patients with traumatic lesions has been undertaken by many investigators from many countries (Russell, Espir, 1961; Conrad, 1947; Luria, 1947; Marie and Foix, 1917). Because of the difficulty of describing exactly the extent of the lesion in these cases, they are generally less useful for establishing brain-behavior relationships. They are, of course, still valuable for clinical syndrome descriptions, and detailed evaluation may provide insights into neuropsychological mechanisms (Luria, 1947).

Brain stimulation studies have been another fruitful avenue. Penfield and Roberts (1959) have reported the greatest experience with this technique. These studies were done in patients with chronic, uncontrollable epilepsy. Results can be summarized as follows: 1) speech output abnormalities occurred with left or right hemisphere stimulation; 2) dysphasia (usually word-finding problems) followed left hemisphere stimulation only; 3) frontal operculum, posterior perisylvian regions and supplementary motor cortex were most clearly related to language functions as determined by stimulation. Ojemann and co-workers have confirmed and extended these observations (Ojemann, Fedio, 1968; Ojemann et al., 1968; Ojemann, Ward, 1971; Ojemann, Blick, Ward, 1971; Ojemann, Whitaker, 1978). 1) Stimulation of left posterior thalamic structures at a low current interferes with verbal short term memory; at higher current true anomia results; 2) the left frontal operculum is universally and specifically involved in language functions; 3) much wider portions of the posterior perisylvian region and of the inferior frontal region are variably involved in language functions. These data are believed to justify three conclusions: 1) lateralization of language function includes the subcortical structures which project to the relevant association cortices of the left hemisphere; 2) between individuals, there is considerable variation in the commitment to language of various brain regions; 3) the extent of language-related cortex is broader than the classical perisylvian map of Déjerine.

One should, return, however, to vascular lesions for the clearest delineation of brain and language relationships. The relevance of the vascular system and the distribution of cortical vessels to dysphasia syndromes was clarified by Foix (1928) (Fig. 8). Subsequent analysis of specific syndromes (as examples, Broca's dysphasia by Mohr *et al.*, 1978; conduction dysphasia by Benson *et al.*, 1973) has refined our understanding of the vascular anatomy. Large scale confirmation of lesion sites in living patients has been most accurately accomplished with computerized tomographic scanning (Naeser, Hayward, 1978) and with radionuclide scanning (Kertesz, McCabe, 1978) of vascular cases. In most of these cases a vascular distribution is implicated. The following is a brief review of the dysphasias and their corresponding vascular pathology. Except as noted, little attempt will be made to account for the evolution of one syndrome into another or for the effects of recovery (natural or treated). Much of this vascular anatomy was alluded to in earlier sections.

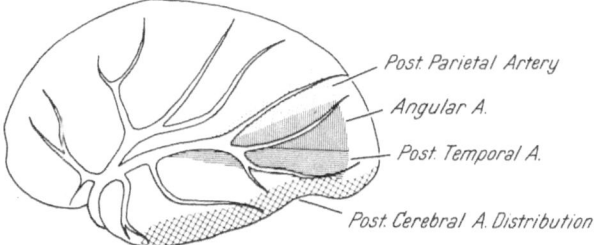

Fig. 8. Schematic rendering of pertinent arterial supply to posterior language regions. Embolic lesions in the distribution of any of the vessels may present as an acute jargon dysphasia. The subsequent course will be different for each. Lasting Wernicke's dysphasia is the result of infarction in the territory of the posterior temporal artery. Alexia with dysgraphia is the outcome of angular artery territory infarctions. Conduction dysphasia will follow posterior parietal artery territory lesions. Posterior cerebral artery distribution lesions will produce pure alexia and/or anomic dysphasia

Infarction in the entire territory of the left middle cerebral artery produces a global dysphasia, often lasting. This extensive lesion may be secondary to an internal carotid occlusion or to an embolus in the origin of the middle cerebral artery. Infarction in the superior division of the middle cerebral artery territory produces a severe nonfluent dysphasia (often global initially) (Fig. 4 A) which evolves into the complex Broca's dysphasia (Mohr *et al.*, 1978; Naeser, Hayward, 1978). Infarction in the frontal opercular branch territory results in early nonfluent dysphasia but with rapid amelioration (Mohr, 1974). Even smaller prerolandic infarction produces an articulatory disorder without lasting language disturbance. Infarction in the parietal operculum (Fig. 5 B) (posterior parietal artery) commonly produces conduction dysphasia (Benson et al., 1973). Infarction in the territory of the angular artery most commonly results in dyslexia with dysgraphia, though larger lesions may cause some auditory comprehension deficits. All of the above represent branch occlusions of the superior division of the middle cerebral artery. The usual cause is probably embolic disease (Caplan et al., 1978).

Infarction of the inferior division of the middle cerebral artery territory produces Wernicke's dysphasia (Fig. 5 A). If only the posterior temporal branch is involved the lesion may be smaller, although a Wernicke's dysphasia is still seen. If more anterior subsylvian branches are affected, pure word deafness or at least a Wernicke's dysphasia with a modality bias against auditory comprehension will result (Hier, Mohr, 1977). These lesions are also usually embolic in origin.

More extensive frontal lesions may result in transcortical motor dysphasia. These lesions may be secondary to a large left frontal infarction sparing Broca's area (Rubens, 1976) or to an infarction in the distribution of the left anterior cerebral artery (Alexander, Schmitt, 1979). The former lesion may be a border-zone infarction secondary to carotid occlusive disease, or it may follow an embolus in a prerolandic branch artery. The latter lesion also involves the supplementary cortex, which may augment the impairment in speech initiation. Smaller infarctions of the supplementary cortex alone will cause a similar disorder (Masdeu et al., 1978). Occlusions either at the origin or in a branch of the anterior cerebral artery are probably embolic.

Extensive parietotemporal lesions may result in a transcortical sensory dysphasia. The vascular basis of this syndrome is uncertain but is presumed to be a posterior borderzone infarction. In fact, the detailed reported cases of this syndrome have not been due to vascular disease.

Posterior cerebral artery occlusion is the classical cause of dyslexia without dysgraphia. Branch occlusions may produce variations in this syndrome, and this is more explicitly reviewed in the chapter on dyslexia. These occlusions are also probably embolic.

This chapter has attempted to review four aspects of the anatomy of language disorders: 1) A localizationist approach was explicitly outlined. Emphasized were the complementary contributions of Déjerine (1914) for demonstrating the basic anterior and posterior language systems, of Geschwind (1965) for delineating the important interhemispheric and intrahemispheric connections of cortical areas involved in language, and of Luria (1966) for expounding the concept of dynamically interrelated centers for different language functions. 2) The anatomical basis for each specific syndrome of dysphasia was reviewed. 3) The roles of different types of brain pathology were considered. 4) The dominant importance of vascular lesions in determining the functional anatomy of dysphasia was stressed.

# Dysphasia with Repetition Disturbance  C

## 1. Broca's Dysphasia

Expressive Dysphasia (Weisenburg and McBride, 1935); Efferent Motor Dysphasia (Luria, 1970); Verbal Dysphasia (Head, 1926); Motor Dysphasia (Goldstein, 1948)

The syndrome of Broca's dysphasia often evolves over a period of months and sometimes years from an initial stage of global dysphasia, but may occasionally appear in its own right as an acute dysphasia. In the completely developed picture of Broca's dysphasia, spontaneous speech is nonfluent and agrammatic, while auditory comprehension is preserved at levels adequate for understanding most conversation and even at nearly normal levels in some patients. Sentence repetition is seriously impaired, although repetition of one- or two-word segments may be possible. Writing is usually on a par with speech. Reading aloud parallels the limited output of spontaneous speech but reading for meaning, though sometimes slow and painstaking, is usually preserved at the level of auditory comprehension. Naming on visual confrontation is often slightly superior to the ability to generate names in spontaneous speech or in response to word definitions. Right hemiparesis affecting the arm more than the leg, buccofacial apraxia, and ideomotor apraxia of the motorically spared left upper extremity are common. In the acute stage, if the initial deficit of auditory comprehension is mild, disturbances of writing are minimal and hemiparesis is not marked, and rapid improvement from a state of severe nonfluency is to be expected over several days or weeks. In these latter patients, agrammatism, as measured in handwriting samples or in a story completion task in which the patient is required to supply the proper grammatical form of a word that completes a sentence, is not prominent.

The patterns of spontaneous speech in patients with Broca's dysphasia vary considerably from patient to patient and over time as recovery takes place. The clinical picture in the individual patient represents a combination of the elements of agrammatism, disordered articulation at the level of phoneme and syllable production, and dysarthria. The overall effect is that of effortful, poorly articulated, hesitating and scanty speech. Rapid shifting from one syllable to another dissimilarly-articulated syllable is very difficult. Phonemic paraphasias occur. Perseverations are common at the syllabic and to a lesser extent at the word level. Melody and rhythm are abnormal (dysprosody), phrases are short and consist mainly of substantives (noun and action verbs) and lack small function words such as prepositions. There is a paucity and simplification of grammatical forms. The absence of small function words has led to the descriptive term "telegraphic speech"; the impoverishment of grammatical forms has led to the label "agrammatic speech". Self-corrections are very prominent and indicate active self-monitoring of errors, a sign of intact auditory comprehension. Repeated failures, despite efforts at self-correction, often result in depression, frustration and anger, with swearing that is usually unintentional. Serial speech (counting, days of the week, alphabet, etc.) and singing are sometimes preserved, even in severely affected patients. Except in the acute period, total mutism is rare.

*Speech Output.* The disturbance at the syllabic and phonemic level of expression may, in some patients, represent the predominant abnormality. Originally described in exquisite detail as the syndrome of phonetic disintegration by Alajouanine, Ombredane and Durand (1939), it has been increasingly referred to in the literature as apraxia of speech or verbal apraxia by many authors who consider it a disturbance in the motor programming of speech sounds (Johns and Darley, 1970; Deal and Darley, 1972; Johns and La Pointe, 1975; Darley, Aronson and Brown, 1975). There are, perhaps, valid objections to the use of the term "apraxia" to designate a disturbance of a linguistic process, namely a disruption of the phonological rules governing selection and ordering of phonemes (Martin, 1974). The disturbance is characterized by impairment of word initiation, phoneme selection, and transitionalizing or blending between phonemes. The pattern of phoneme production is extremely variable and highly context dependent, in contrast to the consistency of distortions of dysarthria. There is a slow and effortful trial and error groping for correct articulatory positions. Incorrect target productions are usually phonemic substitutions, additions, repetitions, or prolongations, in contrast to the consistent distortions and simplifications of uncomplicated dysarthria. Consonant clusters are more difficult than single consonants, which are in turn more difficult to produce than vowels. Initial consonants, especially consonant clusters, and consonants that are embedded in longer and less familiar words, are far more difficult to produce. Errors represent close approximations to target sounds, errors of place being the most common (Trost and Canter, 1974). Single syllables are often broken into two parts. Repetition of target words is more difficult than the production of the same words in spontaneous speech. However, when a target word has been incorrectly articulated, imitation, particularly with direct observation of the speaker's articulatory movements, serves as a model that may improve word production. There is usually a great deal of facial grimacing and struggle behavior with repeated attempts at self-correction. A less obvious component of the articulatory impairment is that of dysarthria, the consistent, non-context-dependent distortion and simplification of phonemes that may be found as an independent entity or mixed with the substitutions and rearrangement of phonemes characteristic of impaired speech output.

The second major component of impaired spontaneous speech is agrammatism. This is characterized by a reduction and simplification of grammatic forms, manifested by the use of short sentences usually restricted to single declarative forms and uninflected verbs. At the most severe level, the patient is restricted to one-word sentences composed chiefly of nouns. There is a loss of the ability to express grammatical relationships, even though there is an availability of words that can be used individually. The agrammatic patient often depends on a stressed or phonologically salient word to begin speech. Small grammatical words cannot be used to open sentences, although they may survive when positioned between two stressed words. This pattern is present on repetition. An example of the difficulty of beginning with the unstressed word is typified by the patient who was trying to answer the question, "What kind of fish did you catch last weekend?" The response was, "Tr - - - too - - - ah - - - lake trout!" The speech patterns in Broca's dysphasia vary from patient to patient along a continuum ranging from severe apraxia of speech with minimal agrammatism to relatively pure agrammatism.

*Auditory Comprehension.* Auditory comprehension in Broca's dysphasia, though appearing normal in ordinary conversation, may break down when studied in more detail. This is particularly true in patients with significant agrammatism. These patients make a significant number of errors when confronted with sentences that are minimally redundant, such as those on the Token Test, particularly the last section which contains reversible sentences (Poeck, Kershensteiner and Hartje, 1972). Two major kinds of comprehension deficit found on formal testing are the inability to follow two- or three-part serial commands (Goodglass, Gleason and Hyde, 1970; Albert, 1972; Heilman, Scholes and Watson, 1976) and the faulty comprehension of sentences in which the full meaning depends on syntactical structure (Caramazza and Zurif, 1976; Caramazza and Berndt, 1978; Heilman and Scholes, 1976; Samuels and Benson, 1979). It has been suggested that the comprehension deficit in Broca's dysphasia parallels the output disorder and that there is an underlying disturbance of a general mechanism serving the syntactical level of both comprehension and speech production (Caramazza and Zurif, 1976). Thus, in Broca's dysphasia, comprehension, though adequate for the highly redundant and context-rich needs of every-day conversation, breaks down when meaning cannot be inferred from word order or other contextual cues that, in ordinary conversation, allow the patient logically to reconstruct the key lexical elements.

*Reading.* Reading in Broca's dysphasia manifests certain characteristic abnormalities. First, evidence is accumulating that reading comprehension deficits in Broca's dysphasia are qualitatively similar to those found in auditory comprehension. Written sentences whose meaning depends on relational and small grammatical words are less well understood than those in which meaning is unambiguously conveyed by the general context and word order of the sentence (Benson, 1977; Samuels and Benson, 1979). Another characteristic feature is the inability of the patient with Broca's dysphasia to name individual letters (literal dyslexia), although the ability to read entire words as single blocks is maintained (Benson, Brown and Tomlinson, 1971; Benson, 1977). This pattern is in marked contrast to patients with the syndrome of dyslexia without dysgraphia who, though unable to read single words, can usually read aloud single letters, and then through a slow, painful process of spelling the word aloud, finally decipher the whole word. There is also difficulty in reading aloud unfamiliar words, nonsense words, and syllables or larger sections of real words while the entire word can be read aloud (Sabauraud, Gagnepain and Sabaraud, 1963). This preservation of global or unit reading of the entire word, coupled with a failure to read individual letters or syllables, is very similar to the dissociation found in Japanese patients with Broca's dysphasia who manifest relatively intact ability to read ideographic Kanji characters and grossly diminished ability to read phonetic Kana characters (Sasanuma, 1975).

*Writing.* Writing performance in Broca's dysphasia usually parallels speech performance. It is characterized by misspellings, letter omissions, perseverations, and agrammatic sentences. Letters are usually large and poorly formed. In severe cases the patient may produce nothing more than an undecipherable scribble; however, even with severe dysgraphia, copying and transliteration are often possible. Arabic numerals are usually better written than letters. When the patient is allowed to spell aloud, use a typewriter, or group anagram letters, performance does not improve. It is common for the first several letters of a word to be correctly spelled, followed by misspellings or failure to complete the word. Mirror reversals of letters, and subsequent self-correction and over-writing are common in patients who are forced to use the left hand because of a right hemiparesis. The finding of relatively spared writing in the acute period indicates an excellent prognosis for overall recovery in Broca's dysphasia.

*Praxis.* Buccofacial and ideomotor dyspraxia of the unparalyzed left upper extremity are quite common in Broca's dysphasia. The phenomenon of buccofacial dyspraxia is strongly correlated with lesions involving the superficial and deep portions of the frontoparietal operculum of the dominant hemisphere, and for that reason, buccofacial dyspraxia is a common accompaniment of Broca's, conduction, and global dysphasia. Since all three of these forms of dysphasia are characterized by a marked phonemic-articulatory disturbance, buccofacial dyspraxia is highly correlated with the presence of phonemic paraphasia. It has been estimated that buccofacial dyspraxia occurs in 90 % of patients with Broca's dysphasia, and in 33 % of those with conduction dysphasia (De Renzi, Pieczuro and Vignolo, 1966). Buccofacial dyspraxia was first described by Jackson (1878), who observed that certain dysphasic patients could not protrude their tongue on command, yet were able to move it purposefully in everyday semi-automatic activity, such as chewing and licking the lips. The phenomenon is characterized by an inability to carry out movements on request involving the facial articulatory musculature, although automatic movements of the same muscles are preserved. These movements include responses to the commands "blow", "stick out your tongue", "cough", "whistle", "sip", "sniff", "puff out your cheeks", "pucker your lips", "lick your lips", "blow", etc. The most susceptible tasks involve coordination of breath stream or phonation with oral movements such as in coughing, blowing, or whistling (Johns and La Pointe, 1975). The ability to close and open the eyes to command is almost always preserved.

Dyspractic responses are often perseveratory in nature, but also include what has been called "verbal overflow", a phenomenon in which the patient repeats the verb instead of actually carrying out the command. Imitation of the examiner's model is not greatly superior to performance when following verbal commands. However, there is a striking dissociation between buccofacial movements evoked through verbal commands and those in response to contextual cues. For example, when the patient is given a real match to blow out or a real flower to smell, performance improves dramatically. Many patients will be unable to mime the action of blowing out a match unless they are allowed to cue themselves by going through the motion of bringing an imaginary match to their lips with their unparalyzed hand. Similarly, the performance of these kinds of actions in everyday life is normal or nearly so. While buccofacial apraxia is commonly found with the dysphasias in which phonemic paraphasia and other articulatory abnormalities are prominent, the fact that buccofacial dyspraxia may occur without significant dysphasia and that Broca's dysphasia occurs without buccofacial dyspraxia lessens the likelihood of a causal relationship (De Renzi, Pieczuro and Vignolo, 1966). Poeck and Kerschensteiner (1975), in their detailed analysis of errors in tests of oral dyspraxia in dysphasic and dyspraxic patients, found the most common errors of movement to be substitution of incorrect facial movements, fragmentary execution of movements, and substitution of verbal movements for nonverbal movements. Perseverative responses were very common. The greatest number of errors in this study were made by patients with Broca's dysphasia, although qualitatively similar but less frequent errors occurred in Wernicke's, amnesic, and global dysphasia.

Mateer and Kimura (1977) demonstrated that while nonfluent dysphasics (primarly Broca's dysphasics) were significantly impaired in their ability to imitate simple single buccofacial movements, both nonfluent and fluent dysphasics fell down in their ability to imitate a series of nonverbal oral movements. In other words, both nonfluent and fluent dysphasic patients were significantly impaired in the imitation of serial facial movements compared to non-dysphasic patients with left or right hemispheric damage. Mateer's more recent work (1978) suggests that the left hemisphere has a specialized capacity for the production of sequenced nonverbal as well as verbal oral movements. In this study, non-dysphasic patients with left hemisphere damage performed at a higher level than dysphasics when producing serial buccofacial movements on imitation, but they did not acquire the tasks as readily as patients with right hemisphere damage or normal controls.

Geschwind (1965, 1975) attributes the failure to perform adequately buccofacial movements in response to verbal commands to a pathological disconnection between an intact auditory verbal processing area (Wernicke's area and the angular gyrus) and an intact cortical buccofacial motor zone. The lesion, situated in the parietal operculum in the case of conduction dysphasia, and in the frontoparietal operculum in the case of global or Broca's dysphasia, isolates the more anterior facial motor zones (the intact left hemisphere region in conduction dysphasia, and the remaining right hemisphere region in Broca's and global dysphasia) from verbal information representing the phonetically and semantically decoded command. According to Geschwind, the important connecting pathway between the posterior language zone and the left premotor and primary sensory-motor facial region is the arcuate fasciculus, while the connecting link between left and right buccofacial premotor zones runs through the anterior corpus callosum.

Ideomotor apraxia of the unparalyzed left upper extremity is also commonly present in Broca's dysphasia, but does not occur as frequently or remain as severe as buccofacial dyspraxia. Although both kinds of dyspraxia are prominent in the acute stage, buccofacial dyspraxia tends to persist to some extent, while limb dyspraxia may disappear after several weeks or months. Limb dyspraxia, however, may occasionally be found without buccofacial dyspraxia in the acute or chronic period. In limb dyspraxia of the ideomotor variety, the patient is unable to carry out commands of the unparalyzed extremity, even though he understands the command. Commands such as requiring the patient to make a first, wave goodbye, salute, beckon someone to come closer, or to pretend to use an object such as a saw or a hammer result in incorrect movements which are usually perseverations, substitutions of one movement for another, or the combination of a previous movement with the required movement. Another type of error is the substitution of a body part for an object (i.e., using the first as a hammer, or using the extended finger as a toothbrush) (Goodglass and Kaplan, 1963). Comprehension of the command can be demonstrated by having the patient indicate which of a number of actions carried out by the examiner is the correct movement. Even then, however, the patient will have difficulty imitating the examiner's model. Yet, as in buccofacial dyspraxia, performance improves markedly when the patient is allowed to manipulate the actual object. Similarly, observation of the patient's performance on the ward or in everyday life provides evidence that he is, in fact, able to manipulate objects successfully when he is not responding to a verbal command.

Liepmann (1905) was the first to call attention to the syndrome of dyspraxia of the left hand in patients with right hemiplegia and severe dysphasia. He labelled the left-sided motor disturbance "sympathetic dyspraxia". Liepmann found this syndrome in fourteen of eighteen patients with right hemiplegia and severe dysphasia, and only six of twenty-three patients with right hemiplegia without dysphasia. The fact that limb dyspraxia sometimes occurs without dysphasia indicates the anatomical and functional independence of the two. Limb dyspraxia is also found in conduction dysphasia, but in this case, because the lesion is posterior to the motor region, there is little or no hemiparesis, and dyspraxia is observable in both upper extremities. Geschwind

(1965, 1975) attributes limb dyspraxia to the failure of a correctly processed and neurally coded spoken command to reach the appropriate motor cortical regions responsible for carrying out the action because of a pathological disconnection between intact posterior language processing zones and remaining cortical motor zones that control the limbs. Ideomotor limb dyspraxia is an important clinical phenomenon for several reasons. First, as is the case with both buccofacial and limb dyspraxia, failure to respond appropriately to verbal command may be confused by the clinician with a failure of comprehension. In Broca's and conduction dysphasia, great care must be taken to distinguish improper response due to dyspraxia from those in which comprehension of the command is defective. Second, the presence of either of these two forms of dyspraxia is a highly reliable sign of a dominant hemisphere lesion.

There is a class of commands that appears to be relatively invulnerable to unilateral left hemisphere lesions. Many patients with severe dyspraxia are able to carry out axial movements to command. These commands include, "Stand up", "turn over", "walk backwards", "sit up", "turn around", "close your eyes", etc. This interesting phenomenon was first noted by Liepmann (1900) and has been commented on in detail by Geschwind (1965). The invulnerability to unilateral language zone lesions of axial or so-called whole body commands extends beyond the phenomenon of dyspraxia to many patients with severe comprehension disturbances. In all probability, therefore, the right hemisphere is involved in their mediation.

*Neuroanatomy.* Mohr and his associates (Mohr, 1976; Mohr *et al.,* 1978) have contributed greatly to the delineation of the syndrome of Broca's dysphasia, its evolution over time, and its patho-anatomical substrate. Based on their own CT scan and autopsy correlates, and an extensive review of the literature, these authors concluded that cerebral infarctions limited to Broca's area and its immediate neighboring structures, including the deep underlying white matter, produce nothing more than a transient speech and language disturbance characterized by the features of so-called verbal dyspraxia with little or no agrammatism. The initial deficit in spontaneous speech ranges from near mutism to a mild dysarthric-phonetic disturbance. In the early stages the speech disturbance is associated with marked buccofacial and limb dyspraxia, dysphasic impairment in writing and minimal impairment of auditory comprehension. Dysphasic features rapidly disappear over days or weeks, leaving behind mild verbal dyspraxia and dysarthria with fully serviceable conversational speech. The full picture of Broca's dysphasia with profound agrammatism occurs with extensive lesions of the frontoparietal operculum and insula that lie within the territory of the superior division of the middle cerebral artery. The initial manifestation is that of global dysphasia, which evolves slowly over months or years to a picture in the chronic stage of persisting Broca's dysphasia. Occasionally the full picture of Broca's dysphasia with agrammatism occurs acutely, but even in these cases the lesion is extensive and exceeds the confines of Broca's area to extend posteriorly into wide areas of the parietal and/or insular cortex.

A conclusion to be reached from these observations is that the right cortical buccofacial motor complex is capable within several weeks of taking over the function of mediating articulated speech. The right cortical articulatory motor complex may perhaps be driven by the still-functioning left posterior language zone, by as yet undefined callosal and subcortical pathways. It should be noted, however, that the final stage of restitution in patients with lesions limited to the region of the posterior left frontal operculum is often marked by persistent non-language dysarthric and dysrhythmic speech features that are rarely found with lesions of the corresponding right opercular region.

Right hemispheric, frontoparietal, suprasylvian lesions, while not producing a significant phonological impairment, have been reported to produce an impairment in the affective qualities of speech by an inability to control the necessary prosodic intonational patterns that convey emotion (Wilson, 1908, 1926; Ross and Mesulam, 1979). Thus, in contrast to the severely affected patient with Broca's dysphasia who may suddenly swear with fluency, the patient with a right suprasylvian frontoparietal lesion may lose emotional speech and yet have no deficit of vocabulary, grammar, or articulation.


## 2. Wernicke's Dysphasia

Sensory Dysphasia (Wernicke, 1874); Receptive Dysphasia (Weisenburg and McBride, 1935); Acoustic Dysphasia (Luria, 1966); Syntactic Dysphasia (Head, 1926)


In Wernicke's dysphasia, spontaneous speech is fluent, paraphasic, and unmonitored, and auditory and written comprehension are poor. Performance in reading, writing, and repetition parallels that of spontaneous speech. The patient is commonly unaware of his speech deficit and produces a flow of uncorrected, meaningless, well-articulated sentences. The syndrome is almost always associated with a lesion of the posterior, superior dominant temporal lobe (Wernicke's area) and because of this, primary motor and sensory signs and symptoms are minimal, although a right hemivisual field deficit is sometimes present. Initial euphoria and gestural hyperactivity, and sometimes, later in the course of the illness, paranoia may accompany the language deficit.

*Speech Output.* At the onset, spontaneous speech may contain phonemic and verbal paraphasias and their combinations, sometimes to the point of neologistic jargon. Later, when some degree of auditory self-monitoring develops, phonemic paraphasias become less prevalent, speech rate becomes somewhat slower because of more frequent self-corrections and word-finding pauses, and verbal paraphasias, circumlocutions, and information-poor words such as "thing" or "place" become more prominent. Regardless of the severity of the dysphasia, there is very little effort in speaking and no dysarthria. Prosody is preserved. It would, therefore, be impossible to detect this form of dysphasia in a patient speaking a language with which the examiner was not familiar.

Augmentations, the addition of extra, unnecessary syllables, words, or phrases to an otherwise adequate response, are common. There is a tendency to continue talking without allowing the examiner to get a word in "edgewise", so-called "press of speech". When speech flow is even more excessive and incessant, the term "logorrhea" is applicable. Continuous, non-corrected paraphasic speech containing little or no information has been called "jargon". Alajouanine, Sabouraud and Ribaucourt (1952) and Alajouanine *et al.* (1964) distinguish between two sometimes co-existent types of jargon: phonemic jargon consisting of strings of words that have been altered in their phonemic composition by phonemic paraphasias, and semantic jargon consisting of a flow of verbal paraphasias connected by incorrect grammatical forms. When the patient produces paraphasic words that are totally meaningless, composed entirely of phonemic paraphasias or a mixture of phonemic and verbal paraphasias, the term "neologistic jargon" is applicable.

In contrast to the speech in Broca's dysphasia, the speech of the patient with Wernicke's dysphasia contains an abundance of small function words and grammatical forms that are, however, used inappropriately. Speech lacks substantives, such as information-conveying nouns, action verbs, or their modifiers. Therefore, despite the profusion of words, very little meaning is conveyed. Characteristically, the patient lacks awareness of his speech errors or even insight into the fact that there is a communication problem of any kind (anosagnosia). The paucity of attempts at self-correction serves as a reliable index of the degree of impairment of auditory comprehension and attention.

In the more chronic stage, some degree of self-monitoring returns but never to the degree found in Broca's or conduction dysphasia. Some patients with Wernicke's dysphasia, particularly those with prominent semantic jargon, detect and attempt to correct only their phonemic paraphasias. Others, mainly those who produce phonemic paraphasias, are considerably annoyed by their verbal paraphasias and are oblivious to their phonemic errors. The latter group shows little disruption of its speech pattern when subjected to delayed auditory feedback (Alajouanine *et al.*, 1964) in marked contrast to normals and many patients with semantic jargon.

The phonemic paraphasia of patients with Wernicke's dysphasia has been studied in some detail (Buckingham and Kertesz, 1975; Alajouanine and Lhermitte, 1973; Green, 1969; Le Cours and Rouillon, 1976; Burns and Canter, 1977). It has been found to consist of complex confusions, alterations, and recombinations of both phonemic and semantic elements. Neologistic jargon is a complex mixture of semantic and phonemic confusions, while semantic jargon consists of strings of well-articulated, unrelated words and paragrammatic forms. It affects both vowels and consonants, and there are frequent transpositions of phonemes both within words and across word boundaries. Complex mixtures of phonemic and semantic alterations are more common in Wernicke's dysphasia, while pure phonemic confusions are more common in conduction dysphasia. According to Burns and Canter (1977), the intrusion of semantically related words and the addition of unnecessary phonemes is more profound in Wernicke's dysphasia than in conduction dysphasia. Errors of place appear to be the most common, and errors involving two or more phonemic feature components are more frequent than errors involving a single component.

*Auditory Comprehension.* Auditory comprehension in Wernicke's dysphasia is severely impaired and generally parallels the degree of abnormality found in spontaneous speech. The patient is often unable to point to the single item in an array of two or three items that has been named by the examiner. Comprehension may be so impaired that, instead of following requests to point or to carry out non-oral motor commands, the patient responds with unsolicited and inappropriate utterances. If the patient can be persuaded to remain silent and direct his attention to the question, auditory comprehension improves noticeably. When the patient is rested and the interview has just begun, comprehension may appear normal. However, after a few questions or commands, the patient rapidly "fatigues" to the point where few spoken requests or questions are understood.

Many patients with Wernicke's dysphasia retain their ability to carry out some whole body or axial commands, even when they demonstrate no other form of auditory comprehension. This is particularly demonstrable when their attention has been gained and they are not simply responding to all questions with an inappropriate flow of speech. It is important, therefore, when assessing auditory comprehension at the bedside, to require more than just a few responses and to include tasks other than whole body commands.

It is generally believed that the failure of comprehension in Wernicke's dysphasia reflects a deficit of processing at the phonemic decoding level, whereas in transcortical sensory dysphasia the phonemic level of auditory processing is preserved but the semantic level, in which the word sound is associated with word meaning, is faulty. A common method of demonstrating a level of auditory processing breakdown is to ask the patient to repeat words and full sentences. While the transcortical sensory dysphasic is able to repeat long, complicated sentences that he does not understand, the patient with Wernicke's dysphasia fails miserably at repetition, even when it can be demonstrated that he is aware he is to repeat and is obviously attempting to do so. Repetition errors consist both of phonemic and verbal paraphasias or a combination of these two forms.

The nature of the phonemic processing deficit has been attributed by many authors, notably Luria (1947, 1970), to an inability to detect phonemically salient cues and to extract the few important phonemic features necessary for identifying and distinguishing meaningful phonemes and words. According to this view, the comprehension deficit is primarily that of a disturbance of phonemic perception. This view is supported by the finding that these patients are least able to match an auditorily presented word to its picture when the possible choices include an array of phonemically similar words (Schuell, Jenkins and Jiménez-Pabón, 1964; Naeser, 1976; Gardner, Albert and Weintraub, 1975). However, several recent studies appear to indicate that phonemic perception in itself plays a minor role, if any, in the comprehension deficit in Wernicke's dysphasia (Blumstein, Baker and Goodglass, 1977; Naeser, 1976).

In a study of same-different phonemic discrimination of dysphasics, Blumstein, Baker and Goodglass (1977) found that while patients with Wernicke's dysphasia make some errors in phonemic discrimination, they make no more than many patients with nonfuent dysphasia who have a lesser degree of comprehension deficit and whose lesions are located above the sylvian fissure. Since the deficit on phonemic discrimination tests does not correlate with the level of comprehension impairment, it is difficult to attribute a causal relationship. It appears, therefore, that the phonemic processing deficit in patients with Wernicke's dysphasia is not at the perceptual or hearing level but at the level in which linguistic significance is associated with adequately perceived phonemes.

Finally, even with a severe comprehension disturbance, many patients with Wernicke's dysphasia perform surprisingly well in certain paralinguistic aspects of comprehension. For example, if the examiner speaks to a patient with either invented, meaningless jargon or even with samples transcribed from a tape of the patient's own voice, the patient's behavior indicates that he is aware of the inappropriateness of the speech. Likewise, it has been shown that these patients are able to distinguish between their own language and other, non-native languages (Boller and Green, 1972).

*Reading.* The comprehension of written material in Wernicke's dysphasia generally parallels auditory comprehension. However, a distinct subgroup of patients whose comprehension of written material is significantly superior to that of auditory comprehension has been identified by Hecaen (1969, 1972). Several detailed case reports describing the relative preservation of reading in this syndrome have been published (Hier and Mohr, 1977; Heilman, Rothi, Campanella, and Wolfson, 1979). It is probable that in this group of patients there is more involvement of the left auditory primary and secondary auditory processing regions than in most Wernicke's dysphasics, and the added element of word deafness results in the greater impairment of comprehension in the auditory than in the visual modality.

*Writing.* Writing in Wernicke's dysphasia is generally on a par with spontaneous speech, although it has been claimed that these patients are more conscious of their written errors than their errors in speech (Hecaen and Angelergues, 1965). Spontaneous writing is often superior to that on dictation, again probably reflecting an element of word deafness. There have been published case reports in which written naming was superior to oral naming (Hier and Mohr, 1977). Because there is no hemiparesis, these patients continue to use the right hand for writing. Natural handwriting is preserved, although written production may be unintelligible. Wernicke's dysphasics may refuse to write but commonly will produce paraphasic, rambling, repetitious texts containing few substantives and concrete action words.

*Neuroanatomy.* The finding in a right-handed patient of fluent, paraphasic, meaningless speech, with few attempts at self-correction and a severe impairment of auditory comprehension, is one of the strongest localizing signs in neurology. Its presence indicates a lesion located in the posterior superior temporal lobe of the left cerebral hemisphere. The lesion may exceed the boundaries of this area and extend into parietal and occipital lobes, but this does not significantly change the quality of the speech and language disturbance. It is possible that the relative sparing of the comprehension of auditory versus visual language (written) reflects relative sparing of the left primary and secondary auditory association regions in the case of the former, and of the inferior parietal (angular gyrus) in the case of the latter (Hier and Mohr, 1977). Concomitant involvement of the parietal region in Wernicke's dysphasia is also reflected by greater difficulty in performing nonlanguage visuospatial tasks, such as copying block designs, and drawing three-dimensional figures. The finding of an unusual amount of phonemic paraphasia implicates additional involvement of the parietal opercular structures.

It is not rare for large lesions of Wernicke's area to fail to produce the full picture of Wernicke's dysphasia, but instead to result in conduction dysphasia (Benson et al., 1973). There is a single case report of a patient with total destruction of Wernicke's area at necropsy in whom no dysphasia was noted during life (Boller, 1973). Another patient with bilateral infarctions, one in Wernicke's area and the other in its right-sided counterpart, developed severe auditory agnosia. Minimal phonemic and verbal paraphasias or other abnormalities of speech output were noted in the chronic period (Oppenheimer, Newcombe, 1978). These cases indicate the importance of individual differences in lateralization of function and of intrahemispheric cerebral organization, as it relates to the posterior language processing area.

## 3. Conduction Dysphasia

Central Dysphasia (Goldstein, 1978); Afferent Motor Dysphasia (Luria, 1966)

In the syndrome of conduction dysphasia, spontaneous speech is fluent but contaminated, to a greater or lesser degree (depending upon severity and time post-onset), by paraphasias that are most often of the phonemic variety. In contrast to Wernicke's dysphasia, auditory comprehension and the comprehension of written material is not noticeably impaired in the non-test setting but, as in Broca's dysphasia, comprehension may break down on more demanding tests in formal investigation. While spontaneous speech may approach normal levels after the acute period, repetition is always impaired, particularly for phrases, sentences, and multisyllabic, unfamiliar and meaningless words. It is the striking discrepancy between impaired repetition and preserved auditory comprehension that distinguishes conduction from Wernicke's dysphasia.

There is almost always anomia, either phonemic paraphasia or total inability to find a word when attempting to name. Reading aloud parallels the performance of repetition, but silent reading for comprehension of text is at the high level of auditory comprehension. Writing usually parallels spontaneous speech in its errors, with errors ranging from minor misspellings, omissions, and reversal of letters to profound paragraphia that mirrors the phonetic errors in spontaneous speech.

Conduction dysphasia most often occurs as a stage of recovery from Wernicke's dysphasia, but may also occur as an acute syndrome in its own right. In the latter case, the prognosis for complete recovery is excellent.

Because the lesion responsible for conduction dysphasia lies at the posterior end of the left sylvian fissure, affecting primarily the cortex and subcortical white matter of the supramarginal gyrus, and often Wernicke's area itself (Green and Howes, 1977), significant paresis is rarely found. A right-sided cortical hemisensory syndrome, sometimes associated with dysesthesias and spontaneous pain similar to that found with the thalamic syndrome, has been reported (Benson et al., 1973). The syndrome of spontaneous pain has been associated with lesions of the parietal operculum (Denny-Brown and Chambers, 1958; Biemond, 1956). Buccofacial apraxia is very prominent, while bilateral limb dyspraxia, also common, is usually less severe and persistent than the dyspraxia of the buccofacial muscles. It has been claimed that dyspraxia is present with conduction dysphasia when the lesion involves the supramarginal gyrus, but not in those cases in which the lesion is limited to the temporal lobe (Benson et al., 1973). Parietal lobe signs, including various components of the Gerstmann syndrome (e.g., acalculia, finger agnosia) and constructional dyspraxia are common accompanying features. Reproduction of spoken or tapped rhythms is commonly defective (Alajouanine and Lhermitte, 1964).

Certain characteristics of spontaneous speech help distinguish conduction dysphasia from Wernicke's dysphasia. In conduction dysphasia, auditory comprehension is spared and the patient is self-critical. For that reason spontaneous speech, while fluent, is often relatively slowed by attempts at self-correction and by word-finding pauses. It is common for a patient dissatisfied with an utterance to attempt by successive, repetitive approximations, to approach and finally reach the target word (conduit d'approche). Although attempts at self-correction may eventually lead to the target word, repeated attempts may, at times, lead the patient further and further away from the intended utterance and result in ludicrous, complex multisyllabic phonemic confusions (conduit d'écart). For example, a patient who intended to say, "Nelson Rockefeller", said, "Nelson Nockenfellen, I mean, Relso Rickenfollow, I mean, Felso Knockerfelson". Patients are usually acutely aware of most of their errors and may respond with anger and frustration, or at times with bemused amazement or even hilarity, at their unexpected utterances. As they become more attuned to their deficit, and as spontaneous recovery occurs, patients tend to anticipate difficulty and to slow the rate of speech in order to avoid certain words and sentences. Many patients learn to pre-monitor their speech and to substitute more easily expressed words and circumlocutory paraphrases. The avoidance of potentially troublesome words and sentences combines with overt selfcorrections and word-finding pauses to slow the rate of speech relative to that found with Wernicke's dysphasia but never to the point of the reduced fluency found with Broca's dysphasia.

It is the act of repetition, where the degree of freedom of word and sentence choice is limited to the model sentence, that paraphasias, usually phonemic but sometimes verbal or combinations of the two, are more prominent. Repetition of single, common words is usually possible, but meaning-less or foreign words, multisyllabic words, and sentences of unusual construc-tion are difficult to repeat. There is greater difficulty with grammatical words than with substantives. Repetition of number series is relatively spared in comparison to other word categories (Geschwind, 1965), and paraphasias in the semantic category of number are almost always of the verbal variety in contrast to phonemic substitutions with other word categories. When verbal paraphasias occur on repetition, they are most commonly perseverations of words previously used in an already repeated sentence, or represent a para-phrasing of the model sentence with close semantic but not phonetic approxi-mation of the intended meaning (e.g., "The child walked away from the large dog" becomes "The boy went away from the big dog").

The anatomical basis of conduction dysphasia has been reviewed in great detail by Benson et al. (1973) and by Green and Howes (1977). Green and Howes were able to find twenty-five cases with post-mortem or surgical confirmation of lesion location. In thirteen cases, the lesion extended from the posterior superior temporal gyrus to the supramarginal gyrus. In nine cases, the temporal lobe was spared and the lesion affected the supramarginal gyrus predominantly. In three cases, the temporal lobe but not the supramarginal gyrus was involved. In another three cases, the lesion was confined to the zone either directly above and/or behind the Sylvian fissure. The following Table 3, modified from Green and Howes (1977, Table A 1, page 151), summarizes the anatomical location and etiology of lesions in twenty-five patients with conduction dysphasia in anatomical confirmation of lesion site and etiology.

It is clear from this table that conduction dysphasia is routinely associated with lesions (mostly vascular) that cap the posterior end of the Sylvian fissure. A significant minority of patients have lesions restricted to Wernicke's area.

The production of conduction dysphasia instead of Wernicke's dysphasia with lesions restricted to the temporal lobe has been accounted for by Kleist (1962) by the presence in these patients of mixed language dominance. In this situation the right temporal lobe is the dominant auditory speech decoding area and the left inferior posterior frontal lobe the dominant encoding area; a lesion of the left temporal lobe would break the connecting link between the two and result in a phonological breakdown in verbal expression without affecting auditory comprehension.

The nature of the defect in conduction dysphasia remains a subject of considerable controversy. It has been considered a variety of sensory dysphasia (Liepmann and Papenheim, 1914), a mild form of word deafness (Kleist, 1962), and a mixture of mild sensory dysphasia and general loss of attention to sensory stimuli (Stengel and Lodge Patch, 1955). The more classical view, represented by the writings of Geschwind (1965) and, more recently, Kinsbourne (1972), is that the failure of repetition stems from faulty trans-mission of accurately decoded auditory information from the posterior speech analyzer to the anterior speech encoding region. Geschwind implicates a lesion of the arcuate fasciculus, a white matter bundle that runs deep to the inferior parietal lobe and connects large portions of the temporo-parietal association

Table 3. *Surgical and post-mortem reports of lesions in twenty-five cases of conduction dysphasia*

| Author | Type of Lesion | H | PT | TGP | ISSM | RSSM | A | PO | INS |
|---|---|---|---|---|---|---|---|---|---|
| Lichtheim, 1884 | infarct | | | M | S | S | | | M |
| Pick, 1898 | infarct | | | S | | | | | M |
| Pershing, 1900 | infarct | | | | M | | M | | |
| Goldstein, 1911 | tumor | | | M | | | | | M |
| Liepmann and Pappenheim, 1914 | infarct | M | | M | S | M | M | | |
| Bonhoeffer, 1923 | infarct | | M | | S | S | M | | |
| Potzl, 1925 | infarct | | | | S | | | | |
| Hilpert, 1930 | abscess | | | | S | S | | | |
| Stengel, 1933 | tumor | | | M | S | | M | | |
| Potzl and Stengel, 1937 | infarct | M | M | | S | S | | | M |
| Goldstein and Marmor, 1938 | infarct | M | | S | M | | M | | M |
| Coenen, 1940 | infarct | M | | M | S | | S | | |
| Stengel and Lodge Patch, 1955 | infarct | | | M | S | S | M | M | |
| Hecaen et al., 1955 | tumor | | | M | S | | M | | |
| Hoeft, 1957 | infarct | | | S | S | | | | |
| Konorski et al., 1961 | tumor | | | S | | | | | |
| Kleist, 1962 | infarct | | S | S | M | | M | | |
| Kleist, 1962 | infarct | M | M | S | S | | | | |
| Caraceni, 1962 | tumor | | | | | S | | | |
| Warrington et al., 1971 | tumor | | | | S | S | | | |
| Warrington et al., 1971 | tumor | | | S | S | S | M | | |
| Brown, 1972 | tumor | | | | | M | S | | |
| Benson et al., 1973 | infarct | | | | S | M | S | | |
| Benson et al., 1973 | infarct | | | | S | S | M | | |
| Benson et al., 1973 | infarct | | M | S | | M | | | M |
| Total cases of partial damage | | 5 | 4 | 7 | 2 | 5 | 9 | 2 | 6 |
| Total cases of severe damage | | 0 | 1 | 8 | 16 | 9 | 3 | 0 | 0 |

Key to location initials: H=Heschl's gyrus; PT=Planum temporale (posterior); TGP=First temporal gyrus (posterior); ISSM=Inferior sylvian supramarginal gyrus; RSSM=Retrosylvian supramarginal gyrus; A=Angular gyrus; PO=Parietal operculum; INS=Insula.

Key to cell letters: M=partial damage; S=severe damage.

cortex with the premotor frontal cortex. Goldstein (1948), on the other hand, believed that a more proper name for the syndrome is central dysphasia, because the symptoms arise not from a failure of conduction of verbal information from sensory to motor regions, but from an impairment of the central core of language, a disturbance of inner speech. Accordingly, even though speech decoding and encoding mechanisms are intact, the artificial and, therefore, more abstract nature of the task of repetition is difficult, while normal spontaneous speech is possible.

Luria (1966) originally viewed the isolated impairment of repetition either as a failure to discriminate heard phonemes or as a form of motor dysphasia in which the kinesthetic or sensory organization of speech motor acts is

6*

impaired. More recently, Luria (Luria and Hutton, 1977) proposed the basic defect to be a failure to attend to the phonological structure of heard words with impairment of conscious analysis of the sound structure of words, rather than their meaning. In this sense, conduction dysphasia is an impairment on the sensory side of speech processing. On the other hand, according to Dubois *et al.* (1964) and Boller and Vignolo (1966), the primary defect is an underlying abnormality of phonemic encoding and, therefore, the deficit lies on the output or expressive side of language processing. Tzortzis and Albert (1974) present evidence that a major factor responsible for the repetition deficit is an impaired memory of heard sound sequences: an inability to retain sound order information without loss of memory for individual sound components.

Warrington and Shallice, in a series of papers (Warrington and Shallice, 1969; Warrington and Shallice, 1972; Shallice and Warrington, 1977), develop a line of evidence that suggests that there are two, often co-occurring, kinds of conduction dysphasia, one associated with the inability to repeat single, relatively infrequent multisyllabic words, which they call reproduction. They believe that the form of conduction dysphasia secondary to reproductive difficulties is characterized by the inability to repeat individual words on request, particularly longer and less frequently used words. The severity of the reproductive deficit would theoretically correlate closely with the severity of phonemic paraphasia in spontaneous speech. In the other form, secondary to repetition difficulty, there is a defect of short-term memory that would not produce abnormalities in spontaneous speech but would give rise to the inability to repeat a series of unconnected, short, familiar words in proper sequence (*e.g.*, a series of digits). They suggest the two forms commonly occur together, but that they represent defects of two independent underlying processes, a verbal motor encoding deficit similar to that proposed by Dubois *et al.* (1964), and a true short-term verbal memory deficit. According to Strub and Gardner (1974), the repetition deficit occurs at a step of auditory verbal processing after phonological analysis of the auditory input has been achieved but before actual encoding of the phonemic components for eventual expression. In some ways this view is similar to that of Luria and Hutton, because it suggests an inability of a damaged auditory processing mechanism to attend both to the semantic and phonemic characteristics of heard speech with a resulting production of irrelevant sound and meaning associations on attempts to repeat, but not during silent listening for meaning.

## 4. Global Dysphasia

In global or total dysphasia all aspects of speech and language are severely affected. Auditory and written comprehension are severely impaired, and speech is grossly nonfluent. Repetition is impossible, a finding that distinguishes global dysphasia from mixed transcortical dysphasia. At best, the patient has at his disposal only a few words or a few stereotyped repetitive utterances. There is no meaningful writing. Even the simple act of pointing to one of three named objects may be performed with inconsistent results. Occasionally, the family reports that the patient appears to understand conversational speech in the home setting and, in fact, the clinician sometimes suspects that the patient understands more than can be demonstrated on formal testing. For example, context-related questions may result in appropriate nonverbal responses, such as crying in response to the question, "Did your family come and see you at the hospital today?", or a gesture of disgust in response to the question, "How do you like the hospital food?". The commands "take off your glasses" and "close your eyes" are almost always responded to appropriately. Other commands, which Geschwind classes as "whole body commands" (Geschwind, 1965) or "axial commands" (Geschwind, personal communication), may also produce appropriate motor responses. These include requests to sit up, turn over, stand up, lie down, straighten up, turn around, look behind you, etc. The reason for the isolated retention of these axial commands is not clear; however, the retained ability to carry them out in the presence of huge left hemispheric convexity lesions suggests a role for the right hemisphere in this phenomenon.

Global dysphasia is most commonly associated with lesions that destroy large portions of the left fronto-temporo-parietal language zone, extending from its anterior-most to its posterior-most poles. In doing so, it crosses the sensorymotor strip, usually extending deeply into the white matter, often to the ventricular surface. The usual etiology, therefore, is occlusion of the main stem of the left middle cerebral artery. However, large deep hypertensive hemorrhages, tumors, and other destructive lesions may produce the syndrome. Neurological signs include right hemiplegia, hemianesthesia, homonymous hemianopia, varying degrees of right-sided sensory neglect, and loss of other intellectual functions beyond that of language. Because of the size of the lesion, the onset of the syndrome is commonly associated with semicoma or severe obtundation.

Numerous studies have shown that when global dysphasia does not improve before the first several weeks following onset, significant long term recovery is not to be expected (Godfrey and Douglass, 1959; Schuell, Jenkins and Jiménez-Pabón, 1964; Sands, Sarno and Shankweiler, 1969; Sarno, Silverman and Sands, 1970). However, several authors (Vignolo, 1964; Mohr et al., 1973; Mohr et al., 1978) have noted late improvement in severe global dysphasics characterized by a significant improvement in comprehension with slow evolution from global to Broca's dysphasia. Mohr et al. (1973) identified a sub-group of patients with global dysphasia who gradually evolve from a state of initial mutism with relative preservation of written naming and matching through a stage of improved oral naming on visual confrontation with continued relative superiority of written naming. Spoken words in the class of picturable nouns are better comprehended than spoken letter names. In a later paper Mohr et al. (1978) offer evidence that the fully developed picture of chronic Broca's dysphasia with prominent agrammatism characteristically evolves over months or years from an initial stage of global dysphasia rather than appearing as an acute and persisting syndrome in its own right.

Sudden but transient global dysphasia may be seen with any large, acute lesion affecting the language zone. Even posterior language zone lesions may produce temporary nonfluency when the lesion is large and sudden in onset. Of particular interest is the unusual and puzzling combination of acute global dysphasia without significant hemiparesis. This is sometimes observed with acute, large ischemic infarcts affecting the left premotor zone and also with hemorrhages involving the deep left frontal lobe anterior to the motor cortex and the posterior limb of the internal capsule (personal observation, A.R.). Associated neurological findings are the presence of a right frontal gaze paresis and minimal right hemiparesis. In this situation the patient suddenly becomes mute and cannot follow verbal commands except for those classified as axial commands. The striking preservation of complex axial commands, sometimes including two- and three-part serial commands, in the presence of severely restricted nonfluent speech and otherwise severely impaired auditory comprehension suggests a large lesion in the cortical or subcortical premotor zone of the left hemisphere. If there is minimal extension inferiorly into the frontal operculum, the syndrome rapidly improves over several weeks through a stage of transcortical motor dysphasia, followed by normalization of speech and language. Acute, large left temporal lobe hemorrhages may also produce profound, but transient, global dysphasia without hemiparesis. The dysphasia evolves over several days or weeks into one of the varieties of fluent dysphasia, in which the major components are impaired naming and reading. Persistent global dysphasia without hemiparesis may also occur with multifocal lesions that spare the sensory-motor region and with dumbbell-shaped perisylvian infarcts which tend to spare the mid- and upper portions of the sensory-motor cortex but destroy the posterior frontal operculum and the posterior temporo-parietal Sylvian region.

Large, deep, left-sided hypertensive hemorrhages involving the basal ganglia often produce a sudden, severe global dysphasia; hemiparesis and hemisensory deficit are profound. Recovery from such deep hemorrhages may slowly proceed over many months to a stage of minimal language impairment. Speech may remain rapid, contaminated by unfinished or telescoped words, and fading or fluctuating volume. Speech output may be relatively unmonitored (to judge by the absence of self-correction and outward concern). It resembles in many respects the nondysphasic speech disturbance referred to as "cluttering" in the speech pathology literature.

# 5. Aphemia

Aphemia is characterized by an isolated loss of the ability to articulate words without loss of the ability to write and to comprehend spoken and written language. Depending on the theoretical point of view of the author, it has been called "pure word dumbness", "anarthria", "pure motor dysphasia", "subcortical motor dysphasia", and "apraxic anarthria". For example, Lichtheim (1885) believed that the combination of preserved language with inability to speak could be explained by a pathological subcortical undercutting of Broca's area that isolated the intact motor speech region by transecting its efferent white matter pathway to brain stem motor centers. Déjerine (1914), on the other hand, was aware that the syndrome occurred with either cortical or subcortical lesions of Broca's area and preferred the term "pure motor dysphasia". Bastian (1887) adopted Broca's original term, "aphemia", which Broca intended to mean loss of the ability to articulate words without impairment of intellect, memory for words or other language functions.

Aphemia is most commonly seen as a residual of Broca's dysphasia, but occasionally occurs as a dramatic initial manifestation of an acute lesion involving the left fronto-central operculum. In the acute stage there is total mutism with or without the ability to phonate. The patient indicates his wish for paper and pencil to answer questions and to make his needs clear. Speech remains severely impaired for several days or weeks and then improves to a stage marked by slow, effortful, scanning and somewhat explosive articulation without passing through a stage of agrammatism, anomia, or other features of dysphasia. Phonemic paraphasias may not appear. A dystonic motor component with a tendency to begin articulatory movements in a state of exaggerated muscular contraction has been identified (Alajouanine, Ombredane and Durand, 1939; Le Cours and Lhermitte, 1976). The inability to achieve and maintain a state of relaxation or intermediate contraction interferes with the production of certain phoneme features such as fricatives and hinders the transition from one phoneme to another. This feature has been described as part of the larger syndrome of phonetic disintegration in Broca's dysphasia (Alajouanine, Ombredane and Durand, 1939). In the final stages of recovery efficient speech is possible, but there is loss of automaticity and there is a need for continuous conscious effort when speaking.

Faciobrachial paralysis, dysphagia, and buccofacial dyspraxia are common at onset. Paralysis of the face may involve the forehead in the acute stage. Upper extremity weakness improves considerably, yet the patient may be left with persistent clumsiness and sensory abnormalities of the hand, particularly of the thumb and index finger. The features of pseudobulbar palsy are lacking. Pathological crying or laughing and other mental and emotional symptoms are not present. Inner speech, as measured by tests of word and picture rhyming, is not affected (Nebes, 1975). Unlike the patient with Broca's dysphasia, the aphemic patient, even during the acute stage of mutism, accurately answers questions and indicates his wishes and needs by writing.

Clinical-pathological studies have shown that aphemia may result either from cortical or subcortical lesions of the left facial sensory-motor region, Broca's area, or both. The well-studied patient of Alajouanine, Pichot and Durand (1949), whose pathological findings were later reported by Le Cours and Lhermitte (1976), developed a sudden complete suppression of speech, free of other language symptoms, that lasted four weeks. In the chronic stage, speech was efficient but was produced in a slowed, syllabical, dysarthric and dysprosodic manner with a tendency towards explosiveness. Post-mortem examination revealed a vascular lesion that involved the cortex and subcortex of the lower three-fifths of the pre-central gyrus and the dorsal-most part of the insula. Other reports with pathologically verified focal lesions include a small vascular lesion of the left frontal operculum immediately subcortical to Broca's area (Bastian, 1887), a small cortical infarct of the surface of Broca's area itself (Déjerine, 1914), a subcortical vascular lesion situated between the insula and putamen (Foix, 1928), and a variety of lesions limited to Broca's area without deep extension (Hecaen and Consoli, 1973). The common pathological finding, therefore, is a lesion, usually vascular, that destroys either the left sensory-motor cortical articulatory region, Broca's area itself, or the subcortical connections between these two areas or their descending outflow pathways. The fact that speech almost always improves to a serviceable level indicates the potential of the homologous right facial-sensory cortical motor region for assuming the function of speech.

# 6. Pure Word Deafness

(Auditory Agnosia for Speech, Verbal Auditory Agnosia)

In pure word deafness, understanding of spoken language is grossly disturbed, yet the ability to speak, read, and write, and, in many cases, to process non-verbal auditory stimuli, remains unaffected. Auditory comprehension and repetition of heard speech are greatly limited by the receptive deficit. The syndrome is "pure" in the sense that it is relatively free from language deficits found with other cerebral syndromes effecting auditory comprehension, *i.e.* Wernicke's dysphasia and transcortical sensory dysphasia. Primary neurological motor and sensory symptoms and signs are not prominent because the neuropathological process is usually restricted to the temporal lobe or lobes.

The term "pure word deafness" (reinen Worttaubheit) was first applied by Kussmaul (1877) to the deficit of patients who could not understand spoken speech but whose own speech and hearing were normal. Later, Lichtheim (1885) described under the term "subcortical (or peripheral commissural) sensory dysphasia" a patient who was unable to understand spoken language, but whose speaking, reading, and writing were intact and who, with the exception of musical sounds, could identify nonverbal stimuli. Lichtheim suggested that the disorder resulted from a disconnection of the left auditory word processing region from both ascending auditory pathways.

In clinical-pathological studies, the usual finding is one of bilateral, somewhat symmetrically placed vascular lesions involving the cortex and the subcortex of the anterior part of both superior temporal gyri, with some degree of sparing of the primary auditory cortex (Heschl's gyrus), more so on the left (see review by Vignolo, 1969 and Ulrich, 1978). There are, however, several well-documented examples of single left temporal lobe lesions apparently isolating Wernicke's area from incoming auditory information by destroying the left auditory radiation and the callosal fibers originating from the opposite auditory region (Liepmann, 1898; Schuster and Taterka, 1926). The proximity of lesions to Wernicke's area explains why pure word deafness is rare; a leftsided lesion must be unusually positioned and circumscribed to isolate from sensory input and yet not destroy Wernicke's area.

Although pure word deafness may result from a variety of etiologies, it most commonly follows bilateral temporal lobe vascular lesions separated in time by months or years (Ulrich, 1978). For that reason, the syndrome frequently occurs either suddenly in a patient who has fully recovered from Wernicke's dysphasia when he suffers a new right temporal lobe infarction, or as a stage in the recovery of a patient with a newly acquired Wernicke's dysphasia. In the latter case, as the paraphasias, writing and reading disturbances disappear, severe impairment of auditory comprehension persists; words and sentences not understood when spoken are easily comprehended when written. At this stage, auditory hallucinations and transient paranoid ideation some-times occur. A peculiar lack of appropriate reaction to painful or other threatening stimuli (asymbolia for pain) has also been reported in some patients (Hemphil and Stengel, 1940).

Unlike other patients with dysphasic comprehension loss, patients with pure word deafness tend carefully to observe the faces of people with whom they are conversing and to use lip-reading as an aid in auditory comprehension. They complain bitterly that although they can still hear, their hearing for speech has been profoundly altered. In this respect, they may resemble patients with bilateral sensori-neural hearing loss whose common complaint is that they can hear what people say but cannot understand them (Olsen and Tillman, 1968). One of our patients complained, "I can hear you talking but it doesn't compute". Other subjective reports include "voice come but no words" (Hemphil and Stengel, 1940); speech is "like foreigners speaking in the distance" (Klein and Harper, 1956); "words come too quickly" and "sound like a foreign language" (Albert and Bear, 1974).

Patients with pure word deafness may be misdiagnosed as hysterical because their auditory comprehension varies markedly from examination to examination. The variation is, however, context-related. Most patients do better when they know in advance the category of the subject under discussion. For that reason, auditory comprehension is superior when the patient intro-duces the subject himself, and it drops drastically when the examiner suddenly changes the topic. Isolated words are less easily identified than words embedded in sentences. Slowing of the presentation rate also improves comprehension.

Conversational speech may contain occasional dysphasic errors, particu-larly word-finding pauses and paraphasias, and may resemble the speech of the peripherally deaf because of poor modulation of pitch and intensity. The tendency to perseverate words and phrases has been noted (Klein and Harper, 1956; Wolfart, Lindgren and Jernelius, 1952).

Although unilateral left-sided hemispheric lesions, particularly those producing Wernicke's dysphasia, may result in a difficulty with matching non-verbal sounds to pictures, the errors are predominantly semantic in nature and are rarely acoustic confusions (Vignolo, 1969). There is little evidence that unilateral left hemisphere temporal lobe disease produces a perceptual-discriminative nonverbal sound recognition disturbance. For that reason the presence of impaired ability to discriminate nonverbal speech sounds in a patient with pure word deafness should add to the suspicion of bilateral disease, even when other neurological signs of bilaterality are lacking.

Albert and Bear (1974) have presented evidence that a basic defect in pure word deafness is slowing of the process of temporal resolution of auditory stimuli. Their patient demonstrated abnormal auditory fusion threshold for clicks and dramatic improvement in auditory comprehension when sentences were presented at slower rates. They suggested that a slowed rate of speech presentation facilitates auditory comprehension by counteracting the adverse effects of pathologically slowed auditory processing. It is equally possible, however, that a slower speech rate gives the patient more time to make educated guesses. Neisser (1967) has observed that the processing of words and their meanings involves an ongoing act of synthesis in which the listener uses knowledge of the world successively to eliminate alternative meanings by making educated guesses based on contextual cues. Thus, Saffran, Marin and Yeni-Komshian (1976) documented in a well-studied case of pure word deafness that either informing the patient of the category of the topic under discussion, indicating the category of words to be identified, or giving the patient a multiple choice array immediately before presentation, significantly aided comprehension. Based on a detailed analysis of their patient's auditory processing errors, they conclude that pure word deafness represents a disturbance of speech perception at the pre-phonetic level, in which auditory material processed acoustically is not accessible for further processing by Wernicke's area. This probably applies specifically to patients in whom nonverbal sound processing is intact, and in whom minimal right hemispheric pathology can be inferred. The finding of a profound left ear advantage on dichotic listening in the patients of Albert and Bear (1974) and Saffran, Marin and Yeni-Komshian (1976) supports this view.

Pure word deafness must be differentiated from other disturbances severely affecting auditory comprehension. These include Wernicke's dysphasia, transcortical sensory dysphasia, and cortical or peripheral deafness. The absence or relative mildness of paraphasia and reading and writing errors distinguishes pure word deafness from Wernicke's dysphasia. Patients with pure word deafness more commonly complain about their auditory disturbance and make obvious attempts to study the faces of people with whom they are conversing for cues. Pure word deafness is distinguished from transcortical sensory dysphasia by the absence of dysphasic symptoms in spontaneous speech, reading and writing, and by the inability of the word-deaf patient to repeat heard speech. In transcortical sensory dysphasia, repetition with poor auditory comprehension is possible because speech sounds are discriminated and perceived normally but comprehension is disturbed at the level of semantic processing. Deafness due to sensory-neural or conduction defects can be ruled out by appropriate audiometric tests. Nonverbal sound recognition must be tested in all patients with word deafness because a defect in this function will otherwise go undetected.

Because there is some confusion in the literature resulting from the inconsistent use of the terms "pure word deafness", "auditory agnosia", and "cortical deafness", a brief discussion of auditory agnosia and cortical deafness follows. Auditory agnosia is sometimes used in a broad sense to mean impaired capacity to recognize both speech and nonspeech sounds and sometimes in a more restricted sense to refer to a selective loss of nonverbal sound recognition.

When authors use the term "auditory agnosia" to refer to the continuum of all auditory recognition problems associated with central auditory dysfunction, the term is usually further divided into auditory sound agnosia, auditory verbal agnosia, and a poorly characterized mixed group. When the more narrow definition is used, auditory agnosia means selective impairment of nonverbal sound recognition; and pure word deafness, selective impairment of the recognition of speech sounds. Although auditory sound agnosia in the absence of a speech recognition disturbance has been reported (Spreen *et al.*, 1965; Albert *et al.*, 1972), it is far more common to find a combination of impaired speech and nonspeech sound recognition. This mixed picture occurs with bilateral temporal lobe lesions, and is often referred to generally as auditory agnosia. It frequently evolves from a state of cortical deafness, and it is often difficult to separate the two entities clinically. Since patients with auditory sound agnosia do not ordinarily complain specifically about their problem, nonverbal sound recognition should be tested in all patients with pure word deafness. This is best done with tape recorded sounds, asking the patient both to name the source of the sound and to match the sound with a picture of its natural source by pointing. A dissociation between defective sound naming and intact sound-picture matching has been reported in word deafness, ostensibly as a result of an auditory callosal disconnection produced by a unilateral left temporal lobe lesion (Denes and Semenza, 1975).

The term "cortical deafness" refers to an extreme lack of awareness of auditory stimuli of any kind with markedly abnormal pure tone audiometric thresholds. It was originally believed that bilateral lesions of the primary auditory cortex resulted in a persistent state of total hearing loss (Henschen, 1918). However, recent animal experiments (Massopoust and Wolin, 1967) and clinical pathological studies on humans (Mahoudeau *et al.*, 1958; Wolfart *et al.*, 1952) leave little doubt that bilateral destruction of the primary auditory cortex produces no substantial permanent loss of audiometric threshold sensitivity. A common clinical presentation is for an otherwise asymptomatic patient with a pre-existing unilateral temporal lobe lesion to become suddenly deaf with the development of a new lesion in the other temporal lobe. After several hours or days hearing returns (normal pure tone audiometric thresholds), but perception of speech and/or nonspeech sounds remains permanently affected (Jerger *et al.*, 1969; Jerger *et al.*, 1972). Occasionally, profound hearing loss persists and the patient may not attend even to very loud noises (Ernest *et al.*, 1977). It is probable that in these instances of profound permanent hearing loss (cortical deafness), bilateral damage extends beyond the primary auditory cortex to include large areas of the auditory temporo-parietal association cortex and/or the auditory thalamus.

# D Dysphasia without Repetition Disturbance

## 1. Transcortical Dysphasias

The features common to the transcortical dysphasias are preserved repetition and cortical damage at or beyond the periphery of the perisylvian language core. The zone of damage is, therefore, related to the more distal territory of the middle cerebral artery or to portions of its vascular border zone with anterior and/or posterior cerebral artery territories. In the case of transcortical motor dysphasia, the lesion may lie totally within the territory of the anterior cerebral artery.

Lichtheim (1885) was the first to call attention to this form of dysphasia. He noted that the preservation of repetition did not fit into Wernicke's previous model (Wernicke, 1874) of motor, sensory and commissural dysphasias (conduction dysphasia, Leitungsaphasie). A lesion affecting either the motor or sensory speech centers or their interconnection would disturb spontaneous speech to the same degree as imitative speech. He hypothesized a pathological separation of the intact speech zone from certain diffusely represented nonlanguage cerebral areas (concept center, Begriffsfeld) that provided motivational and ideational activating influences upon the speech zone during spontaneous volitional speech. He employed the term "commissural" or "white matter pathway dysphasia" (Leitungsaphasie) to indicate this disconnection.

Wernicke (1885—1886) accepted Lichtheim's formulation but preferred to reserve the term "Leitungsaphasie" for what is now called conduction dysphasia. Wernicke coined the term "transcortical dysphasia" to designate those dysphasias with intact repetition. Transcortical dysphasia is now the most generally accepted term that refers to those syndromes in which repetition is well preserved, in spite of considerable loss of spontaneous speech (transcortical motor dysphasia), or severe comprehension disturbance (transcortical sensory dysphasia), or both (mixed transcortical dysphasia).

Many authors have objected to the use of an anatomical label to refer to a disturbance of verbal behavior, while others have proposed other than the transcortical model to explain the retention of repetition in this syndrome. Wernicke himself later expressed reservations concerning the anatomical evidence for the transcortical mechanism, but urged the continued use of the term because of its theoretical usefulness (1908). Bastian (1897) asserted that transcortical dysphasia could result from mild dysfunction of the cortical speech centers themselves. He suggested that subtotal damage would heighten the threshold of excitability so that reaction, though still possible in response to stronger, externally generated stimuli, was no longer possible in response to internally derived volitional stimuli. Goldstein (1948) accepted the notion of partial damage to explain the superiority of repetition sometimes observed during recovery from motor dysphasia, particularly of the post-traumatic variety. Other authors have claimed that even with complete destruction of the motor speech area, repetition may be spared by virtue of the function of the minor hemisphere (Niessl von Mayendorf, 1911). Rubens (1976) and Mohr et al. (1978) (Case 3) reported patients with large lesions of Broca's area and the cortical facial motor complex in whom sentence repetition was possible, probably as a result of the shared functions of the left posterior language and the right articulatory motor regions. Transcortical dysphasia has even been reported as a final stage of recovery after major damage to the central language zone (Gloning, Gloning and Hoff, 1963; Stengel, 1936; Stengel, 1947).

No author has dealt with the question of transcortical dysphasia in greater depth or detail than Kurt Goldstein (1917, 1948). Goldstein proposed the term "isolation of the speech area" for the situation in which Wernicke's area and Broca's area and their interconnections were separated from the posterior "ideational field" by widespread fronto-temporo-parietal pathology. Recently several well-documented, pathologically verified examples of the isolation syndrome have been reported (Geschwind, Quadfasel and Segarra, 1968; Whitaker, 1976), and the term "isolation of the speech area" has reappeared in the literature. By analogy, "anterior isolation" refers to transcortical motor dysphasia, "posterior isolation" to transcortical sensory dysphasia. For years, Luria (1966, 1970) had written about the entity of frontal dynamic dysphasia which appeared to many authors to correspond to transcortical motor dysphasia. In a recent publication, Luria and Hutton (1977) agreed that the two were the same.

From a psycholinguistic point of view, the transcortical dysphasias represent an important opportunity to gain a fuller understanding of the language competence of the isolated perisylvian language zone. From the neurological point of view, transcortical dysphasia, when present in the acute state, suggests complete or partial sparing of the central perisylvian language core (Naeser and Hayward, 1978; Kertesz, Lesk and McCabe, 1977) and thus is helpful in the determination of etiology and prognosis.

## 2. Transcortical Motor Dysphasia

Anterior Isolation Syndrome (Benson and Geschwind, 1971); Dynamic Dysphasia (Luria, 1970)

Transcortical motor dysphasia is characterized by a marked reduction in the amount and complexity of spontaneous speech despite retained ability to repeat sentences, to read aloud, and to name objects presented through a particular sensory modality. Comprehension of written and spoken language is preserved at a high level and is comparable to that found in Broca's dysphasia. Written output matches that of speech. The deficit in spontaneous speech may approach muteness in severe cases. In partially recovered cases, the only residual deficit may be difficulty in generating lists of words belonging to specific letter or semantic categories. The basic defect is in generating a full sentence or string of sentences that supply detailed information in a specific narrative context. The sparing of repetition distinguishes transcortical motor dysphasia from Broca's dysphasia, the syndrome with which it is most readily confused. Lesions responsible for transcortical motor dysphasia are located in the language-dominant frontal lobe, peripheral to Broca's area, either directly anterior or superior to it (Goldstein, 1948; Luria, 1970), or within the mid- and upper premotor zone in the neighborhood of the supplementary motor cortex (Rubens, 1975; Rubens, 1976; Naeser and Haywood, 1978; von Stockert, 1974; Goldstein, 1948; Luria, 1970; Damasio and Kassel, 1978).

Perhaps the most useful bedside method of discriminating transcortical motor dysphasia from other deficits producing reduced spontaneous speech is to compare the ability to construct a sentence or a series of sentences describing common, ordinary occurrences or situations such as the state of the weather with the ability to repeat ready-made descriptive sentences provided by the examiner. For instance, a patient may be unable to describe the weather, except for uttering the word "fine", but will then be able to repeat, "It is a beautiful, warm, sunny day outside". The same patient may struggle to say "fire" after watching the examiner strike a match, yet on repetition can say, "Doctor, you took out a box of matches from your back pocket and lit one" (Rubens, 1975). Even when given specific words, such as "horse" and "carriage", the construction of a full sentence may be impossible (Luria and Tsvetkova, 1968). Overlearned, information-poor, hackneyed expressions and exclamations may be available to the patient, for example, "How do you like that!", "For goodness' sakes!", "Isn't that something!"

Patients with transcortical motor dysphasia may give the impression of having a memory deficit because they often respond that they don't know or cannot remember when asked to discuss their medical history. However, when questions are rephrased so that particular aspects of the history are inquired about specifically, such as, "Where were you when you became ill?", replies are often direct and to the point. There are instances, however, when the patient will not be able to respond to simple questions, except by echolalically incorporating the question into the response. For example, to the question,

"How did you come here today?", the response was, "How did I come here today? I'll tell you how I did come here today. I came here today by . . . for goodness' sakes . . . by taking a . . . by coming here today" (Rubens, 1976). In a recent paper, Luria and Hutton (1977) distinguish between two forms of transcortical motor dysphasia. In one form, which they call perseverative dysphasia, the patient can repeat only a word or a brief sentence and then lapses into perseverative verbal behavior. The inability to produce sequenced utterances is primarily for words rather than for syllables or phonemes, as in Broca's dysphasia. In the second type, Luria's so-called "dynamic dysphasia", patients are free of perseverations and are able to repeat long, complicated sentences.

Lesions responsible for transcortical motor dysphasia cluster around two regions, the inferior posterior frontal and the superior parasagittal frontal zones. When an acute lesion such as a stroke involves the superior premotor zone, there may be total muteness or severe dysphonia for several days. This is followed by the ability to mouth words silently, then to whisper while repeating long, complicated sentences. The hemiparesis involves the leg more than the arm. There may be a contralateral grasp reflex and rigidity of the upper extremity, as well as transient urinary incontinence. Dysphonic features usually disappear after days or weeks, and spontaneous speech improves considerably in one to two months. Patients do not go through a stage of phonemic paraphasia or agrammatism.

Persistent dysphasia, with and without dysphonia, has been reported with left frontal parasagittal tumors (Magnan, 1880; Sweet, 1951; Petit-Dutaillis et al., 1954; Chusid, de Gutiérrez-Mahoney and Margules-Lavergne, 1954; Guidetti, 1957; Alajouanine et al., 1959; Penfield and Roberts, 1959; Arseni and Botez, 1961; Carrieri, 1963; Maroun, Jacob and Gowing, 1970). Persistent dysphonia and mutism have been reported as a postoperative sequel of cerebral commissurotomy (Bogen, 1976). Profound motor dysphasia lasting up to several months may follow surgical excision of the left superior premotor zone (Schwab, 1927; Erickson and Woolsey, 1951; Chusid et al., 1954; Guidetti, 1957; Chavany and Rongerie, 1958; Arseni and Botez, 1961). Schwab emphasized the sparing of repetition with profound expressive dysphasia in fourteen of twenty-one patients after excision of area 6 aB for post-traumatic epilepsy. The cerebral topographical maps of Conrad (1954), Russel and Espir (1961) and Luria (1970) depict numerous left-sided convexity and parasagittal premotor missile injuries resulting in initially severe dysphasia. Luria has remarked on the sparing of repetition in these cases. Dysphasia has also been reported with strokes limited to the territory of the anterior cerebral artery circulation (Liepmann and Maas, 1907; Bonhoeffer, 1914; Critchley, 1930; Hyland, 1933; Cimitri and Victoria, 1937; Poppen, 1939; Petit-Dutaillis, 1956; Guidetti, 1957; Luria and Tsvetkova, 1968; von Stockert, 1974; Rubens, 1975; Damasio and Kassel, 1978). In those case reports with sufficient details of the speech and language impairment, the picture of transcortical motor dysphasia is evident (Petit-Dutaillis et al., 1954; Luria and Tsvetkova, 1968; Liepmann and Maas, 1907; von Stockert, 1974; Rubens, 1975; Damasio and Kassel, 1978). A recently described case with pathologically documented infarction limited to the region of the left supplementary motor cortex was said not to have shown a significant dissociation between repetition and spontaneous speech (Masdeu, Shoene and Funkenstein, 1978).

When the lesion involves the lower premotor zone or partially affects Broca's area, there are mild motor deficits in the form of dysarthria, phonemic errors and occasional agrammatism, and a less striking discrepancy between deficient spontaneous speech and intact repetition. This corresponds to one of the two forms of transcortical motor dysphasia recognized by Goldstein (1915, 1948). This form is compatible with Bastian's theory of heightened excitability threshold due to subtotal or transient damage to Broca's area, and is often found during recovery from penetrating missile wounds. In the second form, articulation is normal but there is a striking lack of speech impulse. The inability to begin speech may be severe enough at times to approach muteness. According to Goldstein, this second form results from a loss of frontal lobe volitional influence on the speech zone produced by a lesion situated midway between Broca's area and the nonlanguage frontal lobe regions.

A disorder that appears to correspond to milder forms of transcortical motor dysphasia has been described by Milner (1964) following partial left frontal lobectomy sparing Broca's area. These patients manifest a distinct reduction of spontaneous speech and tend to answer questions with few words. Although there is no obvious dysphasia in the strict sense, and verbal intelligence is unaffected as measured by standard intelligence tests, these patients have great difficulty with tests of word fluency (Thurstone and Thurstone, 1949). When these patients are asked to write out a list of words beginning with a certain letter of the alphabet, they may look around the room for real objects to name and end up writing very few words. This defect does not follow comparable lesions of the right frontal lobe or of the left anterior temporal lobe, even though lesions of the latter produce measurable disturbances of verbal memory. The reduction of word fluency does not correlate with abnormalities of abstract thinking as measured by the Wisconsin Card Sorting Test. Benton (1968) and Ramier and Hecaen (1970) have confirmed these findings in patients with other types of frontal lobe pathology that spares the language zone. Ramier and Hecaen (1970) have, however, reported minimal but definite impairment of word fluency in patients with right frontal lobe lesions as well, and have concluded that the overall deficit results from a combination of reduced initiation of action associated with frontal lobe damage in general and diminished verbal performance associated with left hemispheric damage in particular.

The older literature also describes similar speech and language impairment with lesions just superior to Broca's area. Marie (1917) reported marked slowness of speech and idea formation with normal auditory comprehension and naming with lesions of the second frontal gyrus just superior to Broca's area. According to Marie, the major difficulty was the condensing of ideas into phrases. Kleist (1935) described under the term "Adynamie der Sprache" a group of patients with lesions of the posterior second frontal gyrus (area 9 of Brodmann) just superior to Broca's area, who, although not dysphasic by conventional testing, manifested reduction of spontaneous speech with difficulty in evoking appropriate words and sentences but with normal articulation.

Because transcortical motor dysphasia is more commonly associated with carotid artery or anterior cerebral artery occlusion than is Broca's dysphasia, a predominantly crural hemiparesis or a brachial paresis with predominance of proximal weakness due to border zone or mesial frontal infarction are more common than in Broca's dysphasia.

In recent years, the relevance to speech and language of the superior premotor zone, particularly the left supplementary motor region and its connections with the neighboring cingulate cortex, has gained attention. Electrical stimulation in the region of the supplementary motor cortex of either hemisphere in man produces either arrest of ongoing speech or initiation of involuntary repetitive vocalization (Brickner, 1940; Erickson and Woolsey, 1951; Penfield and Roberts, 1959; Penfield and Welch, 1951). These responses are similar to the abnormal vocal behavior associated with seizures secondary to epileptogenic tumors situated in the area of the left supplementary motor cortex (Alajouanine, Castaigne, Sabouraud and Contamin, 1951; Arseni and Botez, 1961; Carrieri, 1963; Erickson and Woolsey, 1951; Guidetti, 1957; Petit-Dutaillis, Guiot, Messimy and Bourdillon, 1954; Sweet, 1951). These same tumors are often associated with interictal dysphasia. Cerebral blood flow increases over both frontal parasagittal regions during speech (Larson, Skinhoj and Lassen, 1978). In a recent review, Botez and Barbeau (1971) conclude that the supplementary motor region represents a major cortical structure mediating "... the starting mechanisms of speech ..." and identify the ventrolateral nucleus of the thalamus and peraqueductal gray matter of the mesencephalon as the subcortical elements of the system. According to Sanides (1970), the supplementary motor area represents a paralimbic expansion of the limbic cortex. As such, it may mediate limbic influences on the speech act. In non-human primates, it is impossible to produce significant deficits in vocal responses by extirpating known motor speech areas or to induce vocalization by stimulating these cortical regions. Vocal behavior is affected only when limbic structures are stimulated or lesioned (Myers, 1976; Robinson, 1976). For that reason, it has been suggested that transcortical motor dysphasia, with its sparing of vocal responses evoked through external stimulation (reading aloud, repetition, naming to confrontation) and with its absence of syntactic, semantic or phonemic breakdown, might reflect a loss of limbic influence on the speech act (Brown, 1977) rather than a true dysphasia.

According to Luria and his coworkers, the basic defect in transcortical motor dysphasia is an inability to convert initial thought into sentences to form propositions. Luria and his coworkers (Luria, 1970; Luria and Tsvetkova, 1968) view inner speech as a transitional step between initial thought and final verbal expression. When there is a disturbance of inner speech with its predicative function, a discontinuity between thought and verbal output results. To support this view, they note that patients with this syndrome have more difficulty generating lists of verbs than of nouns, the opposite of what is seen with posterior dysphasia. Even when provided with specific words such as "cart" and "horse", these patients cannot construct a complete sentence. The inability to make sentences from individual words or to describe a simple picture suggests an underlying disturbance of the predicative function of inner speech leading to a defective "linear scheme of the phrase". Both Goldstein and Luria view transcortical motor dysphasia as a breakdown in the relations between language and general nonlanguage thought processes. The question of whether transcortical motor dysphasia represents a true disturbance of language function still awaits further definition.

## 3. Transcortical Sensory Dysphasia

In transcortical sensory dysphasia, auditory comprehension is defective at the level of the linkage of sound to meaning. Phonemic auditory and motor processing are intact and the patient is, therefore, able to repeat words and even long, complex sentences that he does not understand. Spontaneous speech is fluent, empty (loaded with information-poor words such as "place" and "thing"), and circumlocutory. Fluency may be disrupted by word-finding pauses. Phonemic paraphasias are much less frequent than verbal paraphasias. Naming is grossly affected. The ability to repeat sentences without comprehending their meaning distinguishes transcortical sensory dysphasia from Wernicke's dysphasia. When allowance is made for word-finding pauses and for deficient selection of nouns, spontaneous speech is excellent. However, verbal paraphasia may be present to such an extent that speech production is limited to fluent, incoherent semantic jargon. Reading aloud is possible, but understanding of written material is equal to or worse than auditory comprehension. Writing is either more impaired or at the level of spontaneous speech. Echolalia is not present. Hemiparesis and obvious sensory abnormalities are not prominent features.

Comprehension is fragile and context-dependent. It is usually less severely impaired than in Wernicke's dysphasia. An example of the fragility of comprehension with changing context is the performance of a patient who had first responded correctly to several commands such as "I want you to point to your nose". He then responded to the question, "How is your sister?", by saying, "I don't understand. You want me to how is my sister? I don't get that one". On another occasion, after following several commands correctly, he said, "You want me to point to the wall? Wall, wall, I used know that one". Milder examples of transcortical sensory dysphasia appear to correspond to what has been termed "semantic dysphasia" by Head (1926) and "acoustico-amnestic dysphasia" by Luria (1970). There is an impairment of the full meaning of words and phrases and of the overall concept of what has been heard. There is also an intermittent "estrangement of sound from meaning" (Luria, 1970), particularly during periods of fatigue or when the context changes. When the comprehension disturbance is even milder, this syndrome is indistinguishable from anomic dysphasia.

The site of the lesion responsible for transcortical sensory dysphasia has not been well-documented in the literature. Well-studied cases are few. Transcortical sensory dysphasia is often found in combination with other disturbances associated with large lesions of the temporoparieto-occipital region.

These include various elements of the Gerstmann syndrome, dyslexia, constructional dyspraxia, and ideational dyspraxia. The sparing of repetition suggests
relative preservation of Wernicke's area and its anterior connections to motor
speech regions. A lesion involving the posterior pole of Wernicke's area and
its posterior and inferior connections to the parietotemporo-occipital juncture
is, therefore, implicated. According to Goldstein (1948), lesions similarly
placed in the posterior superior temporal lobe may produce Wernicke's dysphasia in one patient and transcortical sensory dysphasia in another. However,
a recent radio-isotope brain scan study by Kertesz (1977) places the lesion
posterior and deep to Wernicke's area. This area encroaches upon the vascular
border zone between middle cerebral and posterior cerebral artery territories
and, in fact, the syndrome is often found with combined posterior vascular
border zone infarctions.

In the clinical setting, the empty, circumlocutory spontaneous speech
without phonemic paraphasias combined with the absence of hemiparesis may
give the impression that the patient is suffering from a nonspecific, general-
ized dementing process, rather than a focal lesion of the temporo-parietal
junction. Transcortical sensory dysphasia may be under-reported for that
reason. It is true, however, that this syndrome may be found as part
of the larger picture of Alzheimer's-senile dementia in its early and mid-stages.
In fact, anomic dysphasia or transcortical sensory dysphasia may constitute
the first symptoms of an early dementia.

# 4. Mixed Transcortical Dysphasia

Isolation of the Speech Area (Goldstein, 1917)

In mixed transcortical dysphasia, meaningful spontaneous speech is scanty or
absent and comprehension is severely impaired. The only verbal activities
are the emission of short, meaningless, repetitive, stereotypic utterances,
the echolalic repetition of the examiner's commands or questions, and the
completion of open-ended sentences supplied by the examiner. The recent
literature contains a number of well-studied cases (Geschwind, Quadfasel and
Segarra, 1968; Heilman, Tucker and Valenstein, 1976; Whitaker, 1976). The
syndrome is found most often with multifocal or diffuse pathology that
involves widespread areas of the anterior and posterior association cortex
and at the same time spares the central perisylvian language core. Etiology
includes border zone vascular lesions secondary to combined stenosis of the
carotid and vertebrobasilar arterial systems, degenerative dementias (particularly Pick's disease), and carbon monoxide poisoning. It may be found with
extensive mesial infarction in the territory of the anterior cerebral artery
(Kyorney, 1975 (Case 1)). The syndrome has occasionally been reported with
left middle cerebral artery territory infarction involving the language zone
(Stengel, 1936; Stengel, 1947).

The full clinical picture is that of a patient who lacks all propositional
speech and evidence of comprehension. When spoken to, sparse, meaningless,

stereotyped, spontaneous utterances are produced, and when questioned, echolalic responses occur. Echolalia has been defined as the unsolicited repetition of verbal stimuli. It is considered by some authors to represent automatic, compulsive, verbal behavior when meaning is not grasped (Brain, 1965; Denny-Brown, 1963; Goldstein, 1948). However, echolalia is occasionally present in patients with transcortical motor dysphasia with good comprehension (Luria, 1970; Rubens, 1976). The completion phenomenon, a compulsive completion of open-ended sentences provided by the examiner, often accompanies echolalia (Stengel, 1964).

Even in the total absence of auditory comprehension, patients with mixed transcortical dysphasia may produce grammatically correct sentences when given incorrect models. "Can you wash himself?" becomes "Can you wash yourself?" (Whitaker, 1976). This type of response has been referred to as mitigated echolalia (Arnaud, 1887), to distinguish it from the complete parrot-like repetition sometimes found in these patients. There is evidence that patients with transcortical motor dysphasia are able to take semantic factors into account when correcting incorrect sentence stimuli; patients with transcortical sensory and with mixed transcortical dysphasia cannot (Davis, Foldi, Gardner and Zurif, 1978). The presence of this phenomenon indicates that the isolated perisylvian language core, though apparently incapable of semantic processing, has the capacity to filter and correct syntactically wrong stimulus sentences.

## 5. Anomic Dysphasia

Nominal Dysphasia (Head, 1926); Amnesic Dysphasia (Goldstein, 1948; Weisenburg and McBride, 1935)

Anomia, the inability to generate names on visual confrontation and in spontaneous speech, is a symptom common to all forms of dysphasia. Only when the severity of the naming deficit stands out in comparison to the relative mildness of the other language deficits is the term "anomic dysphasia" applicable. In anomic dysphasia there is difficulty finding words in spoken and written forms of expression, while comprehension of spoken and written language, reading aloud, and repetition are spared. The question of whether anomic dysphasia exists in its own right as an isolated loss of memory for words, independent of any other language or thought disturbance, has been much debated.

The isolated loss of the ability to produce names was first described by Kussmaul (1874) and Broadbent (1884) under the term "amnesic dysphasia". Pitres (1898) was the first to regard it as a syndrome in its own right, characterized by a loss of memory for words independent of other language abnormalities and of specific sensory modality. Head (1926) later introduced the term "nominal dysphasia" to refer to the same syndrome, and considered it part of a more general disturbance of symbolic representation. Weisenburg and McBride (1935), after attempting to group their patients into one of three syndromes, expressive dysphasia, receptive dysphasia, and a mixed expressive-receptive group, were left with a small group of patients who defied classification because of an isolated inability to recall words that was disproportionately severe when compared to the other expressive and sensory language abnormalities. They therefore formed a fourth category, amnesic dysphasia, that designated an isolated, independent language deficit for the recall of words. Goldstein (1924, 1948) attributed amnesic dysphasia to the loss of abstract attitude found with brain damage in general but made manifest by pathology of the language zone. He pointed to the well-known variability of the naming deficit, in which words not available in one context appear in another. Goldstein maintained that it is not the memory of words that is lost, but rather the categorical thought processes necessary for the realization of the full concept of the sought-after word as a collection of abstract attributes. Other authors have denied the independence of anomic dysphasia as a pure syndrome and have viewed it as the residual of a more general dysphasic disturbance (Déjerine, 1914; Marie, 1906; Marie and Foix, 1917).

The major feature in the spontaneous speech of patients with anomic dysphasia is that of word-finding difficulty in the context of fluent, grammatically well-formed sentences. The inability to find words, predominantly nouns, results in circumlocutions, word-finding pauses, and the production of non-specific words such as "place", "thing", or "thing-amabob". The patient may be able to use a word as a verb—even as he is trying to recall it as a noun (eg., "I don't know what you call it, but you comb your hair with it"). Naming of objects on visual presentation is generally on a par with the ability to generate names in spontaneous speech. Naming failure on sensory presentation tends to involve all modalities equally (Goodglass, 1968; Spreen et al., 1966), in contrast to the naming failures of the agnosic patient, which are unimodal and improve if material is presented through other sensory channels. Possible exceptions are the so-called modality-specific anomias, such as optic dysphasia, that are sometimes regarded as naming disturbances limited to a particular sensory modality (see below). In anomic dysphasia, the naming of pictures is no more difficult than that of real objects (Corlew and Nation, 1975; Hatfield and Howard, 1977); in visual object agnosia, picture identification is typically more difficult than the identification of real objects (Rubens, 1979). These dissociations serve as markers for the presence of agnosia, or as some authors suggest, as evidence for a disconnection between specific sensory regions and the language zone (see below). Naming performance in anomic dysphasia is strongly related to the factor of word frequency. Words used commonly in the language (Newcombe, Oldfield and Wingfield, 1964; Wepman et al., 1956), and words used frequently because of daily contact with particular objects (Fraisse, Noizet and Flament, 1962), are more available to the anomic patient than less common words. There is a subgroup of patients, however, who have difficulty with common words and who tend to produce low-frequency alternatives when naming (Wepman et al., 1956).

Anomia is the least useful localizing sign in dysphasia. It is commonly present in diffuse or multifocal brain damage, and may occur as an early symptom in degenerative dementia, the most common form of which is Alzheimer's disease. Often the first or only language symptom in brain tumor, regardless of localization, is word-finding difficulty. Botez (1962) felt that the presence of verbal paraphasia in naming difficulty was relatively reliable for localizing the tumor to the temporo-occipital region, but that in the absence of verbal paraphasia and circumlocution, anomia was an unreliable localizing sign in brain tumor. Kertesz and McCabe (1977) have recently reconfirmed the fact that Broca's, conduction and Wernicke's dysphasias, when due to acute vascular lesions, often evolve over time into anomic dysphasia. The most commonly reported focal site of damage in anomic dysphasia is the left temporal or temporo-parietal region (Pitres, 1898; Nielson, 1947; Brain, 1961; Alajouanine et al., 1957; Newcombe et al., 1971). Mixtures of anomic dysphasia with Gerstmann's syndrome, dyslexia, and constructional dyspraxia are common with angular gyrus lesions, while anomic dysphasia occurs in pure form or with dyslexia when the lesion involves the inferior temporo-occipital convexity.

Benson (1978) has recently proposed a clinical classification of anomia that includes word production anomia, word selection anomia, semantic anomia, category-specific anomia, and modality-specific anomia. In word production anomia, the patient fails to produce the correct word even though indicating that he knows it. The required word is replaced by phonemic paraphasia or failure to respond at all. Phonemic or semantic cueing is often of great help. This type of naming deficit occurs most commonly with lesions involving the frontal and/or centro-parietal operculum, and is part of the picture of Broca's or conduction dysphasia. In Benson's word selection anomia, the patient admits that he is unable to remember the appropriate word, but may claim that it is "on the tip of his tongue". He is able to provide an adequate functional description or other circumlocutory identification. Phonemic cueing is less effective, and semantic cueing is usually of no help, but the patient readily chooses the correct words from a group offered by the examiner. This form is found with lesions of the inferior temporo-occipital junction, within or directly superior to area 37 of Brodmann. In semantic anomia, there is an associated inability to accept the appropriate word when it is offered by the examiner. Verbal paraphasias, circumlocutions and absence of response are common. This type of anomia may be associated with Wernicke's dysphasia and transcortical sensory dysphasia, which corresponds in many respects with the semantic dysphasia of Head (1926) and the acoustico-amnestic dysphasia of Luria (1966). The responsible lesion involves the temporo-parietal junction.

The last two forms are less common. In category-specific anomia, the patient is unable to name items belonging to a specific category, such as colors or body parts, while others are named adequately. The most common example is color anomia, which is usually associated with the syndrome of alexia without dysgraphia and is considered by some (Geschwind and Fusillo, 1966) to represent an agnosia for colors. The patient is unable to name colors shown him or to point to a color that is named by the examiner, yet performs normally on tests of color perception. There are at least two varieties of this disturbance (Oxbury et al., 1969). One is a visual modality-specific disorder typically associated with alexia without dysgraphia and considered to result from a disconnection of an intact right visual area from an intact left language zone (Geschwind and Fusillo, 1966; Oxbury et al., (Case 1), 1969). In this disorder the patient is able to respond correctly to visual-visual tasks by appropriately coloring ling drawings with crayons and to verbal-verbal tasks ("What is the color of a banana?"). The second variety is considered to be part of a general dysphasic disorder arising from damage to the left parietal lobe. These patients cannot produce visual-visual and verbal-verbal responses (Oxbury et al., (Case 2), 1969; Kinsbourne and Warrington, 1964; De Renzi et al., 1972).

Finally, there are the modality-specific anomias, the best known of which is optic dysphasia. The term "optic dysphasia" was introduced by Freund (1889) to describe one of his patients with a left parieto-occipital tumor that had produced a homonymous hemianopia and dysphasia. The patient's naming ability was impaired primarily for visually presented objects. The case report is incomplete and therefore of little value, except for Freund's hypothesis that the dissociated visual naming deficit resulted from a right occipital lobe-to-left speech area disconnection. Currently, optic dysphasia refers to the condition in which a naming deficit is limited to the visual modality, and the patient shows that he recognizes the objects that he has failed to name either by indicating their use or by pointing to them when they are named. Many authors consider optic dysphasia to be a form of visual object agnosia.

In "tactile dysphasia" or tactile anomia, the patient fails to name objects palpated with the left hand, but then can, with the left hand, tactilely match objects to their identical mates. Tactile anomia has been reported with pathological or surgical lesions of the corpus callosum and is thought to represent a tacto-verbal disconnection (Geschwind and Kaplan, 1962; Gazzaniga and Sperry, 1967; Rubens, 1975 (Case 2)). Auditory anomia, inability in an otherwise nondysphasic patient to name a nonverbal sound with retained ability to point to its matching picture, has recently been reported by Denes and Semenza (1975).

# 6. Dysphasia from Left Subcortical Lesions

Speech and language disorders in cases of thalamic lesions have been described for many years (Fisher, 1959) but the recognition of a specific language syndrome has only occurred with recent reports (Mohr et al., 1975; Reynolds et al., 1979; Alexander and LoVerme, 1980). The availability of CT scanning has made the diagnosis of thalamic lesions more reliable and has allowed diagnosis of small lesions which would previously have gone unrecognized. Recognition of all cases of thalamic hemorrhage, for example, has allowed more accurate elucidation of the clinical features (Walshe et al., 1977) and of the dysphasia associated with such lesions.

Although earlier reports of language abnormalities with thalamic hemorrhage mention many of the basic elements, Mohr et al. (1975) were the first to provide a clear clinical picture of the dysphasic syndrome. Their patient was initially stuporous with unintelligible, mumbling jargon speech. Over several days she fluctuated between periods of alertness with near normal language to periods of drowsy inattention with hypophonic speech and logorrheic, paraphasic language. Two months later language was normal. Subsequent clinical reports (Sameral et al., 1976; Reynolds et al., 1978, 1979) largely confirm this clinical picture of fluent language with paraphasic spontaneous output but relatively good auditory comprehension and intact repetition. Dysnomia and dysgraphia are common, but dyspraxia and dyslexia are not. Speech is variable with periods of hypophonia, exaggerated response latencies and mumbling, interrupting basically normal speech. Attention is often poor, and right body neglect may be prominent. Prognosis may be relatively favorable.

Experimental and clinical evidence from other thalamic disease provides some evidence that these deficits have specificity, at least for subcortical structures. Patients with thalamic tumors are often dysphasic (Smyth and Stern, 1938; Cheek and Taveras, 1966). Diminished voice volume, unintelligible mumbling and bizarre responses are noted. Inattention, perseveration, lack of insight and memory loss are also prominent.

Thalamotomies performed for the treatment of movement disorders have provided thousands of cases of limited thalamic lesions (Riklan and Levita, 1969; Selby, 1967; Bell, 1968). Speech deficits or dysphasia occur only in patients with left-sided thalamotomy but are particularly common in patients with bilateral lesions. Dysarthria and hypophonia are the most frequent findings. Dysphasia is less common (17 %), although some word finding problems may be seen acutely in up to 42 % of patients with left thalamotomy (Selby, 1967). The language disorder is usually limited to an anomia which improves.

The thalamic stimulation studies undertaken by Ojemann and colleagues (1968 a, 1969 b, 1971 a, 1971 b) confirm that the left thalamus is involved in speech mechanisms. Stimulation in the posterior thalamic nuclei produces impaired verbal short-term memory or impaired object naming, depending upon the current strength used. These effects are not seen in all patients undergoing stimulation, and alternative hypotheses about the nature of the stimulation effect have been offered (Van Buren, 1975).

Thalamic infarction does not often provide a suitable lesion for analysis of language deficits. Infarction is usually secondary to hypertensive vascular disease. The small lacunar infarcts of the thalamus are not associated with dysphasia (Fisher, 1965). Infarction secondary to posterior cerebral artery occlusion will produce thalamic damage, but the temporal lobe is involved in these cases as well. Any naming or memory deficits may be attributed to temporal lobe damage (Benson et al., 1974).

In comparison to thalamic hemorrhage, there is very little information about the effects on language of putaminal hemorrhages. A single series of patients with CT-diagnosed putaminal hemorrhages provides evidence that dysphasia can occur (Hier et al., 1977). The quality of language disorder is variable, but is apparently similar to dysphasia following thalamic hemorrhage.

Cases of left thalamic hemorrhage or putaminal hemorrhage with dysphasia seen on our service were recently reviewed (Alexander, LoVerme, 1980). The language disorder was qualitatively similar in most. About 50 % showed fluent, paraphasic speech with nearly intact comprehension, mild anomia and normal repetition. Reading comprehension was near normal, but writing was impaired. Praxis was normal. Attention was normal. Affect and insight were blunted. The long-term outcome was good. Another one-third of the patients had similar dysphasia, but the impairment was greater. Speech was softer and episodes of mumbling occurred, but language was still fluent and paraphasic. Naming was poor, but repetition remained near normal. Paraphasias were more prominent and often of an extended English jargon variety (for example, a harmonica was called a small penny arcade orchestra). Comprehension was impaired as a result of attentional problems, but also at higher level linguistic tasks and on sequential tasks. Writing was severely impaired. Some of the patients in this group had mild ideomotor apraxia. Attentional deficits, right neglect, lack of concern and perseveration were common. Long-term outcome was less favorable. Approximately one-fifth of the patients had a lasting Wernicke's dysphasia.

The severity of the language syndrome described above was not critically dependent on lesion site. Within the group, patients with more severe dysphasia tended to have large anterolateral, that is, putaminal, hemorrhages while more mildly dysphasic patients had small, posteromedial, that is, thalamic, hemorrhages. There were exceptions to this dissociation, however, and no definitive correlation between deficit and hemorrhage locus could be made. A larger series of patients may confirm the suggestion that putaminal lesions are more significant for language dysfunction.

Patients with left thalamic hemorrhage or putaminal hemorrhage will have other helpful clinical signs. If seen early, there is nuchal rigidity. Patients with thalamic hemorrhage may show signs of midbrain compression (coma, miosis, Parinaud's syndrome with downward directed gaze). Patients with putaminal hemorrhage will have a severe right hemiplegia and much less sensory loss. Patients with thalamic hemorrhage will have severe hemisensory loss and less hemiparesis. Many patients with thalamic hemorrhage will eventually develop a central pain syndrome.

The similarity of the language disturbance in this syndrome to that in the traditional transcortical dysphasias has been noted. In both disorders, repetition and automatic speech are remarkably well preserved. Cappa and Vignolo (1979) have reported three patients with thalamic hemorrhage in whom detailed language testing emphasized the transcortical features. The preserved repetition is believed to imply that phonological elements of language are preserved. The dysphasia seems to be an impairment at the semantic levels of language (Cappa and Vignolo, 1979).

Current understanding of this syndrome is limited, but associated neuropsychological deficits seem to be involved in production of the total clinical picture. Attentional deficits are probably due to disruption of nonspecific arousal pathways in the deep periventricular regions. Unilateral neglect is a specific unilateral attentional deficit; such a disturbance has been clearly demonstrated in nondominant thalamic hemorrhage (Watson and Heilman,

1979). Memory disorder may be specifically a result of thalamic damage (Victor *et al.*, 1971). The language disorder is somewhat harder to analyze. The effects of an acute deep mass lesion on intracerebral circulation may produce patchy, diffuse ischemia in a borderzone pattern; one would expect a transcortical dysphasia from this mechanism. Expanding, destructive deep lesions may disconnect cortical language areas from subcortical structures and disrupt afferent and efferent systems. It is not evident how this mechanism could result in transcortical features, but this hypothesis might explain the observation of global dysphasia following subcortical lesions (Naeser *et al.*, 1979). If posterior ascending systems are disrupted, one might see the Wernicke's dysphasia as found in some subcortical hemorrhage cases. In fact, careful analysis of the exact anatomy of subcortical lesions (hemorrhage or infarction) has provided evidence that ascending and descending fiber systems from the internal capsule (medial geniculate to superior temporal Heschl's gyrus, for example), may be disconnected from the perisylvian language zone. These anterior and posterior disconnections may cause the articulation disturbance and the language comprehension impairment respectively (Naeser *et al.*, 1979). Finally, damage to subcortical nuclei (thalamus, putamen, etc.) might be critical for the emergence of dysphasia. Available data from other types of subcortical pathology (summarized above) provide evidence only for naming disturbances and dysarthria as definite sequelae of subcortical nuclear lesions.

Subcortical hemorrhages produce a characteristic speech and language disorder in most cases. There is tentative evidence for differences in severity and long-term prognosis depending upon a thalamic or putaminal focus and depending upon the amount of extension into white matter tracts. The neuropsychological mechanisms of this language syndrome are not known, though coincident disorders in attention and memory may blur the margins of the language disorder. Although these deep subcortical hemorrhages produce a dysphasia similar to that of lesions of the frontal and parietal borderzones (transcortical), specific neurolinguistic information which might clarify this similarity is not yet available.

# Disturbances of Reading and Writing E

A review of early accounts of dysphasia (Benton and Joynt, 1960) credits Valerus Maximus in 30 A.D. with the earliest description of acquired dyslexia: a case of head injury producing an isolated acquired dyslexia. After an interval of almost two millenia, sporadic reports of acquired dyslexia became more common in the 1800's but definitive steps toward understanding the acquired disorders of written language were not made until Dejerine (1891, 1892). His descriptions of the clinical findings and anatomical correlations of acquired dyslexia, with and without dysgraphia, remain essentially unaltered as the foundation of the study of disorders of written language.

Reading and writing are complex activities which involve various motor systems, visual perceptual systems, and linguistic and symbolic skills. Disruption of any of these elements could potentially produce a deficit in written language. In this section we shall consider only acquired disorders of reading and writing in persons with previously normal written language. Developmental deficits in reading and writing acquisition (developmental dyslexias) constitute a related but different cluster of clinical and neuropsychological problems. The elaboration of distinctive clinical syndromes (Mattis *et al.*, 1975, for example) and the determination of the critical pathologic anatomy (Galaburda and Kemper, 1979) or physiology of developmental language disorders are not well established.

To review acquired dyslexia and dysgraphia, several arbitrary divisions must be made. Although there is clearly a close relationship between reading and writing and between written language and spoken language, for purposes of discussion certain distinctions will be made. We consider these syndromes as 1) *disorders of written language associated with dysphasia;* 2) *disorders of reading and writing in the absence of dysphasia;* 3) *disorders of reading alone;* 4) *disorders of writing alone.*

# 1. Dyslexias

## Disorders of Written Language Associated with Dysphasia: Dysphasic Dyslexia

The disorder of written language in dysphasic patients is generally similar to that of spoken language. Patients with fluent dysphasia are typically free of hemiparesis, write with their dominant hand and produce written language comparable to their spoken output: variously paraphasic, paragrammatic and empty of content. Their reading comprehension parallels their auditory comprehension. Patients with nonfluent dysphasia are typically hemiparetic, write with their nondominant hand and produce effortful, agrammatic and limited writing (which is not simply secondary to use of the nonpreferred hand). In fact, any major discrepancies between written and spoken language suggest that the patient suffers from a restricted disorder of one input or output modality rather than a primary disturbance in language. Differential impairments in written and spoken language are the clinical basis for diagnosis of the "pure" syndromes: pure word muteness (aphemia), pure word deafness, pure word blindness (pure dyslexia) and pure dysgraphia.

Some apparent discrepancies in written and spoken language may be seen if the examination is limited to reading aloud. The most important example of this difficulty is in conduction dysphasia. These patients may read aloud with hesitations, omissions, and especially literal paraphasic errors while their reading comprehension is quite good (Benson *et al.*, 1973). In the transcortical dysphasias reading aloud is not always indicative of the level of retained written language skills. Additional specific aspects of written language in the various dysphasic syndromes follows.

*Broca's dysphasia:* Nielsen (1938) considered the reading deficit in anterior dysphasias to be one of the "unanswered problems of dysphasia", and his analysis of autopsy studied cases failed to advance a solution. More recent investigation has revealed these patients to have deficits in auditory and reading comprehension which are qualitatively similar (Gardner *et al.*, 1975; Zurif *et al.*, 1976; Benson, 1977; Samuels and Benson, 1979). They have an impairment in using grammatical and relational words to guide them to comprehension, relying instead on the overall context and on the implications of word order. They have particular difficulty understanding 1) lengthy material; 2) abstract, nonpicturable items; 3) information contained in elaborate syntactic form. Reading aloud imposes additional constraints. These patients have particular difficulty naming (reading aloud) individual letters (literal alexia) (Benson *et al.*, 1971). They have similar difficulty reading small, grammatical words, even when those words are phonetically identical to short nouns or numbers (Gardner and Zurif, 1975).

Writing in Broca's dysphasia is similar to speech. It is slow and effortful and focuses on a few substantive elements for carrying a message. Individual words are often deformed by misspellings, letter omissions and reversals and perseveration. Longer attempts are agrammatic. These deficits are not a simple reflection of dominant hand paresis or limb dyspraxia, as there is no improvement with block anagrams or typing. Writing arabic numerals may be preserved, and copying print is relatively accurate (Hecaen *et al.*, 1963).

Relative preservation of written language in Broca's dysphasia may indicate a good prognosis. Alternatively, recent anatomical studies of Broca's dysphasia highlight the large frontal and parietal lesions in lasting nonfluent, agrammatic dysphasia (Naeser, Hayward, 1978; Mohr *et al.*, 1978). Persistent reading and writing problems of Broca's dysphasia may be secondary to this more extensive language area lesion; the relatively good prognosis for Broca's dysphasia with some preserved writing may be an epiphenomenon of less extensive frontoparietal brain damage.

*Wernicke's dysphasia:* Most patients with Wernicke's dysphasia have impairments of written language which are similar to spoken. Patients with severe comprehension deficits have less relative superiority for comprehending substantive words, and they may benefit from lengthy and redundant messages (Gardner, Denes, Zurif, 1975). Their writing is as paraphasic (paragraphic) as their speech, and shows the same word-finding impairments for substantive nouns and action verbs. Using their preferred hand, the script and orthography may be well preserved despite the lack of content.

While the majority of patients with Wernicke's dysphasia have comparable deficits in written and spoken language, groups which have relatively preserved written language or relatively preserved spoken language can be isolated (Hecaen *et al.*, 1968; Hecaen, 1972). Confirmation comes from a series of cases with anatomical information (Mohr *et al.*, 1978), suggesting that more anteriorly placed temporal lobe lesions will allow relatively preserved reading comprehension, while more posteriorly placed temporo-parietal lesions will produce relatively impaired reading. As these anatomical discrepancies increase, the resulting language abnormality will more closely resemble one of the single modality disorders, pure word deafness with limited temporal lobe lesions or dyslexia with dysgraphia with limited parietal lobe lesions. In a similar manner, the more anterior (more word-deaf) Wernicke's dysphasic will have relatively better writing spontaneously than to dictation. This relative preservation may even allow some naming through the written modality alone (Hier and Mohr, 1977). The more posterior (more dyslexic) Wernicke's dysphasic may have some preserved writing to dictation.

As all three language disorders (pure word deafness, Wernicke's dysphasia and dyslexia with dysgraphia) may result from limited infarction in the territory of the inferior division of the middle cerebral artery, small differences in size or placement of this lesion may have major consequences for language. Furthermore, many of these relative differences may only emerge with time. It is not unusual for patients with an acute Wernicke's dysphasia to evolve to a chronic disorder relatively limited to a reading comprehension or an auditory comprehension deficit. With some unusual etiologies, such as herpes simplex encephalitis, exceptionally severe damage to the temporal lobes may account for the prominence of word deafness (Hier and Mohr, 1977).

*Conduction dysphasia:* As noted above, the major reading deficit in these patients is in reading aloud, which may resemble their paraphasic speech and repetition. This occurs despite reading comprehension at least as good as auditory comprehension. Review of autopsied cases of conduction dysphasia did uncover instances of impaired reading comprehension, but this is unusual (Benson *et al.*, 1973). Writing is typically abnormal with omissions, substitutions and misspellings in a fluent framework. Writing generally parallels the speech abnormality but is often profoundly disturbed.

Conduction dysphasia has been described with lesions in a variety of locations clustered around the posterior end of the sylvian fissure (Green and Howes, 1977). Benson and Geschwind (1969) have speculated that the lesions with the more posterior parietal involvement will have the denser deficits in writing and reading comprehension.

*Transcortical dysphasias:* Written language in these disorders has been investigated less definitively, but case reports demonstrate generally comparable impairments in written and spoken language (Rubens, 1976; Alexander, Schmitt, 1979). Reading aloud is preserved in transcortical motor dysphasia; the deficits seen in Broca's dysphasia are not present. Reading comprehension is usually preserved in transcortical motor dysphasia, although difficulty with complex, sequential material may follow the large left frontal lesion. In transcortical sensory dysphasia reading comprehension is severely impaired, although some accurate reading aloud can occur. In fact, this syndrome blends into the more posterior categories of Wernicke's dysphasia (Hecaen *et al.*, 1968; Hecaen, 1972) described above.

Writing deficits in transcortical motor dysphasia are comparable to speech. Some short responsive writing may be possible, but longer productions and spontaneous narrative output are not. Many of these patients are not hemiparetic, but use of the right hand may be limited by a grasp reflex. In transcortical sensory dysphasia writing is of paraphasic fluent form and may improve somewhat to dictation, as described previously for the more posterior Wernicke's dysphasia.

The pathologic anatomy of these syndromes is described in earlier chapters dealing with the language disorders, as well as in the section on Wernicke's dysphasia above. The posterior transcortical syndrome with damage in the parietal-temporal-occipital border-zone is clearly related to the posterior type of Wernicke's dysphasia. Small differences in site and placement of lesions posterior to Wernicke's area may result in various forms of language disorder including dyslexia with dysgraphia, anomic dysphasia and transcortical sensory dysphasia. As a result of the posterior lesions involved however, the last two disorders also may be associated with considerable reading impairment (Benson and Geschwind, 1969).

## Dyslexia with Dysgraphia in the Absence of Dysphasia

Independent of any significant disturbance in speech or auditory comprehension, there can be impairment of reading and writing. This neurobehavioral dissociation marks the syndrome of dyslexia with dysgraphia. In 1891 Dejerine described a patient with severe dyslexia and dysgraphia but with only minimal dysphasia. Postmortem examination revealed an infarct limited to the left angular gyrus cortex with penetration into the underlying white matter (Fig. 4). Subsequent clinico-pathological reports confirmed the relationship between the dominant hemisphere angular gyrus and acquired dyslexia with dysgraphia (Nielsen and Raney, 1938).

The most common etiology of this syndrome is a small infarction in the angular gyrus. This type of limited infarct in the distribution of a cortical branch of the middle cerebral artery is presumed to be embolic in most cases. Dyslexia with dysgraphia has been described in cases of tumors, arteriovenous malformations and trauma as well. Whatever the basis, associated abnormalities on neurological examination may be negligible. A right hemianopia or partial right hemianopia is found in some patients. With sufficient parietal involvement a deficit in right body cortical sensation may occur. Hemiparesis is unusual.

The dyslexia in these cases is almost always severe. Benson and Geschwind (1969) state that most patients will have dyslexia for letters, words and numbers. Hecaen and Albert (1978) believe that most cases have some retention of letter reading ability. There is agreement that functional reading comprehension is severely decreased and that facilitating mechanisms are not helpful. Tactile facilitation (finger tracing of letters) and letter-by-letter mental assembly of words are without benefit. Finally, these patients are unable to understand words spelled to them. Benson and Geschwind (1969) interpret these deficits to be parallel disturbances in "reading" through each sensory modality: visual (print), tactile (tracing) and auditory (spelling). This multi-modality loss of letter and word recognition is a characteristic of dyslexia *with* dysgraphia. Patients with dyslexia *without* dysgraphia have a recognition loss limited to the visual modality.

Dysgraphia in this syndrome is usually as severe as the dyslexia. The patients can usually form letters but rarely manage to assemble words. The dysgraphia is present with spontaneous writing and with writing to dictation. There is also difficulty copying written material.

These patients often have associated neuropsychological deficits. By definition any oral language disorder must be minimal or the patient would have a fluent dysphasia with dyslexia; but a mild anomia is very common in dyslexia with dysgraphia. This observation does not preclude dyslexia with dysgraphia emerging with time from a posterior type Wernicke's dysphasia.

In addition to anomia, most of these patients will have one or more of the following: 1) right-left disorientation, 2) finger agnosia (often a naming difficulty), 3) dyscalculia and 4) constructional dyspraxia. The first three of these signs plus dysgraphia constitute the Gerstmann syndrome (Gerstmann, 1930). The calculation disturbance is often most marked and may even extend to impaired number reading and recognition. The spelling recognition disorder described above is mirrored by an inability to spell aloud even very common words.

While various interpretations of this bi-directional spelling loss have been essayed, the most parsimonious (although also unproven) interpretation has been suggested by Benson and Geschwind (1969) and Geschwind (1965): patients with angular gyrus lesions have lost the ability to form associations across sensory modalities. The various deficits described above (multimodal dyslexia, severe dysgraphia, bidirectional spelling impairment and, variously, dyscalculia, right-left orientation, finger agnosia and perhaps constructional dyspraxia) have in common a dependence on crossmodal associations. Whatever the neuropsychological mechanism(s) that links these deficits with dominant hemisphere angular gyrus lesions, it should be clear that this syndrome of dyslexia with dysgraphia is not simply a more severe form of dyslexia without dysgraphia.

There are two recent case reports of patients with dyslexia with dysgraphia for whom the reading comprehension level was critically dependent upon the method of testing (Albert et al., 1973; Mohr, 1976). These cases can help elucidate additional issues in dyslexia. In both reports, discrepant findings between the testing of comprehension with traditional methods and the testing with experimental methods suggested an additional clinical type of dyslexia. The patient of Albert et al. (1973) had a large left retrorolandic glioma partially resected. At the time of testing he had a dense right homonymous hemianopia and mild right sensorimotor signs. Language testing revealed anomic dysphasia, dysgraphia and a dense dyslexia, but his spelling and recognition of spelled words were relatively preserved. Memory and constructions were abnormal; most of the elements of the Gerstmann's syndrome were present. Despite his dense dyslexia the authors were able to show considerable reading comprehension when context, auditory or picture cues, or semantic categorization were utilized. He was able to recognize nonverbal visual symbols. He could sort letters into their place in the alphabet and recognize the correct written form of words. This combination of deficits is not accounted for by the concept of an angular gyrus lesion described above. The authors review a similar case of Dejerine and Thomas (1904), in which a similar cluster of findings was described. The tentative explanation of these cases was a disconnection of angular gyrus from frontal motor centers bilaterally, combined with a disconnection of visual input from the angular gyrus. The latter produced dyslexia; the former produced dysgraphia. The deeply-placed, extensive subcortical lesion was able to produce two disconnections, one interhemispheric and the other intrahemispheric.

Mohr (1976) described another patient with a deep left posterior lesion (probably demyelinating) with a similar collection of apparently discrepant findings: dyslexia and dysgraphia but good two-way spelling with preserved reading comprehension in a test mode of matching written words to dictated words. To explain this finding, a disconnection hypothesis similar to that of Albert et al. (1973) is offered.

Both of these case reports emphasize key aspects of reading comprehension: 1) it is not a unitary disorder; a broad range of tests might demonstrate retained functions; 2) preservation of oral spelling may indicate preservation of angular cortex; 3) extensive deep lesions may be the substrate of this somewhat different dyslexia; 4) a simple two-way visual-verbal disconnection does not seem to allow for the discrepancies in this case.

Additional observations about these cases should also be made: 1) both patients had unusual etiologies (multiple sclerosis with probably multifocal disease, and widespread recurrent glioma) which are primarily white matter processes and whose disruptive margins cannot be determined; 2) the visual-verbal disconnection may apply for some avenues of reading comprehension (the phonetic verbal-visual route) but not for others (iconic, for example); 3) this level of reading comprehension may be facilitated by efferent programming of the sensory area following verbal or pictorial cueing (Caplan, 1978); 4) the nondominant hemisphere may possess some capacity for iconic and semantic comprehension when tested properly (Gazzaniga, Hillyard, 1971; Heilman et al., 1979).

Additional pieces of evidence suggest that the nondominant hemisphere may be involved in some semantic comprehension. Patients with angular gyral cortex lesions may eventually develop semantically related paralexia despite unchanged severe dyslexia (Benson, Geschwind, 1969). In addition, patients with dyslexia without dysgraphia may demonstrate similar semantic comprehension (Kreindler, Ionasescu, 1961). In this broader context, it is reasonable that slowly developing posterior subcortical lesions (glioma, multiple sclerosis) might produce dyslexia without dysgraphia, but over a time course which allows some semantic comprehension to emerge in the right hemisphere, particularly when potential responses are cued. The dysgraphia is a second, unrelated consequence of one or more of a number of factors: multiple lesions, mild paresis, dyspraxia or even a second disconnection.

## Dyslexia without Dysgraphia (Pure Dyslexia, Agnostic Dyslexia)

In 1892 Dejerine described a patient with the remarkable finding of severe dyslexia with fully retained ability to write. Subsequent postmortem examination revealed infarction of the left medial occipital lobe and of the splenium of the corpus callosum. Dejerine proposed that all visual input to the left hemisphere angular gyrus was lost as a consequence of this pair of lesions. The undamaged left parietal region accounted for preserved writing and spelling aloud. Since that report numerous examples of this clinical syndrome have been published, and there has been general agreement about the pathologic anatomy (Fig. 4). This syndrome is considered one of the most dramatic examples of the interhemispheric disconnection syndromes (Geschwind, 1965).

Foix and Hillemand (1925) provided further delineation of the anatomy of this deficit. They emphasized that dyslexia without dysgraphia occurred with infarction in the distribution of the left posterior cerebral artery while the spoken language disorders resulted from left middle cerebral artery territory infarction. Since that time investigations and clinical reports have focused on several residual issues: 1) the nature of the commonly associated color agnosia and the reasons for its unpredictable occurrence; 2) the critical importance of the lesion in the splenium; 3) the range of left parieto-occipital lesions other than total posterior cerebral artery territory infarction which might produce this syndrome; 4) the neuropsychological mechanism of the reading loss.

As this syndrome usually follows infarction in the left posterior cerebral artery distribution, associated neurological deficits can be anticipated. Most patients will have a right hemianopia. Optokinetic nystagmus may be reduced on the side of the lesion because of involvement of the posterior corpus callosum (Ling, Gay, 1968). With infarction in the entire territory of the left posterior cerebral artery including the thalamus, a marked right-sided hemisensory loss will occur. Involvement of medial temporal structures will result in severe verbal memory disturbance (Benson, Marsden, Meadows, 1974). Rarely, a right hemiparesis secondary to peduncular infarction may be seen (Benson, Tomlinson, 1971).

Typically, patients with dyslexia without dysgraphia have a severe dyslexia for words (verbal dyslexia) with relatively good reading of letters and numbers (Benson, Geschwind, 1969). Some patients are able to read words such as "Coca-Cola" which have familiar configurations (Ajax, 1967). The severity of the reading loss is variable, and cases with loss of letter and number reading have been described (Caplan, Hedley-White, 1974). The patients who are able to read individual letters may be able to read in a letter-by-letter manner, spelling and sounding the word as they progress. It is asserted that many patients are able to read somesthetically by tracing letters with a finger (Ajax, 1967) although again they would have to piece together the word letter-by-letter (Benson, Geschwind, 1976). Through either system reading would be laborious and inefficient.

Writing, by definition, is not disturbed. Nevertheless, some minor errors in orthography may be seen, and because the patient is unable to read his own written productions he cannot rely on rereading his output to assist in maintenance of a narrative train. Despite these mild difficulties, this syndrome is dramatically characterized by the discrepancy between the ease of normal narrative writing and the inability then to read the very same writing.

Spelling and recognition of spelled words are normal. The normal writing and normal spelling represent preserved function of the intact dominant angular gyrus. The syndrome of dyslexia without dysgraphia is not so much a disorder of written language as it is a deficit in the visual modality of language input. In these ways, it differs from syndromes with lesions of the angular gyrus in which a true disorder of language, albeit limited to the written form, is seen.

Associated neuropsychological deficits are common. Verbal memory impairment may follow infarction of the medial temporal region. A mild anomia in speech is common, perhaps secondary to extension of the infarct into inferior temporal areas (Benson, Marsden, Meadows, 1974). Copying of writing and drawing is usually abnormal but, if done slowly and slavishly, may be fairly accurate. Written calculation is usually abnormal. Visual agnosias for objects and colors may be found. Visual object agnosia is usually transient (Caplan, Hedley-White, 1974). Lasting visual object agnosia in these cases is probably a result of additional right occipito-temporal damage (Benson, Segarra, Albert, 1974). A lasting color agnosia is more common and occurs in about 70 % of the patients with this syndrome. The abnormality in color recognition is strictly limited to visual-verbal associations (Geschwind, 1965). The patients have no complaints about their color vision, and colors are used appropriately. Subtle tests of hue discrimination are normal, and the isochromatic color blindness cards are read normally. They are unable to name presented colors or to select a named color from an array. These characteristics distinguish this disorder from acquired cortical color blindness following bilateral inferior temporal-occipital lesions (Pearlman *et al.*, 1979).

Several hypothetical mechanisms for this two-way color naming deficit of some patients with dyslexia without dysgraphia have been suggested. Geschwind (1965) believed that colors arouse only visual-verbal associations, limiting the access to color names to the direct route to the left hemisphere language area. Stachowiak and Poeck (1976) offered a linguistically based explanation of the difference between typically impaired color naming and typically preserved object naming. They suggested that in a confrontation naming task, object names are retrieved as nouns with considerable lexical flexibility while color names are retrieved as adjectives (a more difficult task) with no lexical alternatives.

Since the excellent review of dyslexia by Benson and Geschwind (1969) additional clinical reports have confronted this issue, as well as additional, primarily anatomical, issues in dyslexia without dysgraphia. Cummings *et al.* (1970) described a patient with dyslexia without dysgraphia but with preserved color naming. The dorsal splenium was partially intact, suggesting a possible anatomic basis for the dissociation. Subsequently, a number of reports with radiologic or postmortem anatomical information have confirmed this relationship (Ajax *et al.*, 1977; Vincent *et al.*, 1977; Johanssen and Fahlgren, 1979). In cases of vascular origin, the dorsal splenium may be spared because it lies in the arterial borderzone between the anterior and posterior cerebral arteries (Johanssen and Fahlgren, 1979), and there is postmortem evidence that just this dorsal preservation may be seen in cases with intact color naming (Ajax *et al.*, 1977). Several tumor cases without color naming deficit have had a similar ventral or ventrolateral placement (Greenblatt, 1973; Vincent *et al.*, 1977). Other reported cases with preserved color naming have had a lesion limited to the ventrolateral left occipital cortex and subcortical white matter, disrupting the ventral efferents of the visual association cortex (Ajax, 1967). This implies that the major pathway from right visual association cortex to the left angular gyrus is through the left visual association cortex, at least for verbal material.

The presence of a right visual field deficit does not identify those with or without color agnosia. Some patients have had a complete hemianopia, some a partial or transient visual field disturbance, and others none. The critical anatomical features of pure dyslexia without color agnosia may be either 1) preservation of the dorsal splenium in patients with a right hemianopia or 2) a lesion of the left ventral occipital (fusiform and lingual gyrus) region in those cases without a hemianopia. The recent report of Johansson and Fahlgren (1979) provides further delineation of this anatomical relationship. Three patients with ventrolateral occipital embolic infarcts had no hemianopia and spared color naming; a patient with a dorsomedial occipital infarct had a right hemianopia and color agnosia. Their review of the vascular anatomy of the occipital lobe demonstrates that these two occipital regions are supplied by different branches of the posterior cerebral artery.

The external sagittal striatum is the probable pathway from left ventral occipital regions to angular gyrus. Small lesions disrupting that pathway might again produce dyslexia without dysgraphia, and depending on lesion site a right hemianopia might or might not be seen. While this limited subcortical lesion is hard to find in isolation, there are at least two cases which may be examples of this subangular disconnection of visual-verbal pathways (Greenblatt, 1976; Stroka *et al.*, 1973). This localization is very similar to the two cases of unusual dyslexia with dysgraphia described above (Albert *et al.*, 1973; Mohr, 1976). The absence of dysgraphia in the subangular cases of Greenblatt and Stroka *et al.* could be a consequence of smaller lesions.

Albert *et al.* (1973) and Mohr (1976) have described retention of semantic reading comprehension in patients who were otherwise severely dyslexic *with* dysgraphia. Possible mechanisms for this were listed in the previous section. When similar tests have been given to patients with dyslexia *without* dysgraphia, similar uncovering of retained comprehension is seen. Kreindler and Ionasescu (1961) demonstrated considerable reading comprehension at the letter, word and sentence level by asking for a match of written material to spoken targets. This phenomenon of auditory facilitation or unblocking was effective while being performed; no lasting effect in reading was seen. Subsequently, these matching tests and others have demonstrated considerable retaining reading comprehension in patients with dyslexia without dysgraphia. This retention has been observed in subangular cases (Staller *et al.*, 1978) as well as the traditional splenio-occipital cases (Stachowiak, Poeck, 1976; Caplan, Hedley-White, 1974) with right hemianopia and color agnosia.

A brief summary of the anatomy of dyslexia without dysgraphia might be as follows. This syndrome represents disconnection of visual-verbal pathways from occipital association cortex to the left angular gyrus. This disconnection occurs most commonly from infarction in the left medial occipital cortex and the splenium of the corpus callosum, but a variety of lesions of diverse etiology may disrupt the transfer of visual-verbal information anywhere along the course from left visual association cortex to angular gyrus. Color agnosia is a result of disconnection of visual pathways through the dorsal splenium if there is left calcarine destruction.

This schema for posterior reading disorders has implications for the mechanisms of dyslexia. As traditionally formulated, a symmetrical two-way visual-verbal disconnection does not seem to account for some of the data on dyslexia: 1) the effects of cueing or unblocking (Caplan, Hedley-White, 1974; Mohr, 1976); 2) the breadth of retained semantic comprehension (Albert *et al.*, 1973); 3) the dissociation between color naming and reading. The observations of Levine and Calvanio (1978) suggest that the dyslexia *without* dysgraphia patient has a deficit in visual perceptual analysis of compound related items (such as letters). This deficit is primarily visual and may represent the functional result of the classical disconnection lesion. The effects of cueing and the semantic comprehension may represent alternative pathways from language to vision (Stachowiak and Poeck, 1976), efferent alerting of visual systems through thalamic connections (Caplan and Hedley-White, 1974) or non-dominant hemisphere semantic functions (Hecaen and Albert, 1978).

Although its physiological mechanisms are not known and its exact anatomies are not completely determined, for pure dyslexia the concept of intrahemispheric and interhemispheric disconnections remains the best explanation of the acquired disorders of written language.

*Treatment* of acquired dyslexia is not often reported. The course of natural recovery is generally felt to be poor. Benson and Geschwind (1969) were not optimistic about the treated or untreated outcome. Ajax (1967) described the fruitless outcome of two years of daily treatment in a young man with pure dyslexia. His treatment consisted of reading a text while listening to it read from a tape. Ajax concluded that the prognosis for fluent reading was poor, but, as there is no primary language deficit, a well-motivated person might be able to circumvent the disability in reading. Reports of other patients followed for long periods of time with no improvement (Goldstein *et al.*, 1971) echo the history of Dejerine's original patient (1891), whose deficit remained stable. Johansson and Fahlgren (1979) stated that their patients with ventrolateral occipital infarcts had a better prognosis for reading disability than those with medial occipital infarcts. These patients with the better prognosis might be identified by normal color naming, minimal hemianopia and normal optokinetic nystagmus. Hecaen *et al.* (1952) described 7 patients with left occipital lobectomies with sparing of the corpus callosum. Postoperative dyslexia without dysgraphia abated in 6 months, but reading remained difficult. Preservation of the ventral splenium interhemispheric connections may provide more possibility of recovery. There are, however, several reported examples of this pattern of injury with a poor outcome for reading (Ajax, 1967). Establishing a reliable natural history and prognosis for the various forms of dyslexia clearly is needed as a first step toward identifying treatment candidates.

Many suggested treatments are based on procedures for teaching reading to children (Goldstein, 1948) or on remedial reading procedures (Wepman, 1951), but there is little evidence that they are widely helpful. A technique of multiple oral rereading of extended text (borrowed from remedial reading methology) was used to increase reading efficiency in a single patient with pure dyslexia whose deficit had not responded to a program of single word recognition drills (Moyer, 1979).

Implications for treatment may be present in the unblocking techniques described above. Beginning with Kreindler and Ionasescu (1961), many investigators have demonstrated retained comprehension if the avenue of access is properly selected: auditory-visual matching, for example. Unfortunately, none of these studies has demonstrated any carryover of reading comprehension to the natural setting of silent reading of text. Neurolinguistic factors (contextual semantic constraints) can be manipulated to control performance in reading aloud (Saffran, Schwartz, Marin, 1976), but there is no evidence that this facilitates reading for comprehension in any other setting.

Many patients with pure alexia are able to read in a letter-by-letter fashion through tactile or auditory (reading letters aloud) modalities, but Benson and

Geschwind (1976) have pointed out that this may remain inefficient and inadequate. Practice does not seem appreciably to improve the usefulness of this method.

Dyslexia is not a unitary disorder of written comprehension. Analyzed in clinico-anatomical terms, through neuropsychological performances or through neurolinguistic parameters, several different forms of reading deterioration will emerge. Therapy for these individual forms should be individualized based on the patient's overall retained neurological, neuropsychological and neurolinguistic capacities.

## 2. Dysgraphias

Writing requires the complex interaction of mechanisms of motor control, or praxis, and of visuospatial and kinesthetic integration in addition to the symbolic basis of a language system. Furthermore, writing is not a highly practiced skill such as speech. Therefore, it is not surprising that writing abilities should be fragile and that brain dysfunction of numerous sorts might disrupt normal writing. A pure and isolated disturbance of writing after brain damage is uncommon, and several identifiable categories of writing disturbance can be found associated with other neuropsychological deficits. Dysphasic dysgraphia has already been described for the various types of language disorder in the section on dyslexia. Other important categories of dysgraphia are spatial dysgraphia, dyspraxic dysgraphia, callosal dysgraphia, pure dysgraphia and dysgraphia with acute confusional state.

*Spatial dysgraphia* is a disorder of writing following nondominant hemisphere lesions resulting in visuospatial impairment (Hecaen and Albert, 1978). It may be clearly differentiated from dysgraphias of dominant hemisphere syndromes by its association with other signs of the primary visuospatial perceptual disorder in calculations (number alignment), reading (left neglect or pseudo-alexia) and drawing. These signs of nondominant hemisphere dysfunction are usually most prominent with retrorolandic lesions (Hecaen *et al.*, 1956).

The writing disturbance in large nondominant hemisphere lesions is characterized by 1) a wide left hand margin of the paper; 2) reiteration of all elements in the script: letters, individual slashes or loops of letters or even syllables; 3) deviation of the writing upward from the horizontal (Hecaen and Marcie, 1974). Each of these elements has a clear relationship to a more basic deficit in visuospatial processing: 1) left-sided neglect; 2) perseveration of disrupted spatial elements, similar to overdrawing on constructional tasks and 3) inability to recognize relative angulation of lines. Despite these spatial disruptions, linguistic elements of syntax, meaning and word selection will be normal.

*Dyspraxic dysgraphia* is a form of disordered writing following dominant hemisphere lesions. The relevance of disturbances of learned movement (dyspraxia) to writing disorders should need no emphasis (Henschen, 1922). Dyspraxic dysgraphia is found with other evidence of dominant parietal lobe disease: ideomotor dyspraxia, constructional dyspraxia, anomic dysphasia, impaired spelling and mild dyslexia. A larger or more ventral parietal lesion would produce dyslexia with dysgraphia; and, in fact, the writing disorders of dyspraxic dysgraphia and parietal dyslexic dysgraphia are similar.

The writing disturbance is said to have a dyspraxic quality (Hecaen, Albert, 1978) because of apparent loss of smooth, automatic production of written elements. Letters may be inverted or reversed and perseveration of elements may occur. The disordered graphics will be seen with spontaneous writing or writing to dictation; copying may be somewhat better but is still abnormal. The use of alphabet blocks will circumvent some of the grapheme production errors.

While there are dyspraxic qualities to this form of dysgraphia, there are other factors which contribute to the abnormal written production. Abnormal spelling and word retrieval deficits (anomia) will also be seen in the graphics of this disorder. Valenstein and Heilman (1979) described dyspraxic dysgraphia in a patient with no language difficulty. Reading, spelling and naming were all normal. There was, however, severe limb and buccofacial dyspraxia. In this case, the disturbance in writing was apparently a pure reflection of dyspraxia.

Limb dyspraxia and dysgraphia are commonly seen together as related disorders in patients with *callosal dysgraphia* (Geschwind, 1965). Following surgical section or vascular infarction of the corpus callosum, the nondominant hand will be dyspraxic and dysgraphic (Geschwind and Kaplan, 1962). These deficits are related to the left hand dyspraxia and dysgraphia of right-handed patients with Broca's dysphasia; in Broca's dysphasia the interhemispheric disconnection occurs in the left hemisphere rather than in the corpus callosum (Geschwind, 1975). Both of these syndromes have left sided dyspraxia and dysgraphia because motor control of the left hand will be disconnected from the left brain language zones. In the unusual event that cerebral dominance for handedness is not in the same hemisphere as for language, then disconnection dyspraxia and dysgraphia will follow a lesion in the hemisphere dominant for handedness in the complete absence of dysphasia (Heilman *et al.*, 1973; Heilman *et al.*, 1974). It is not clear, however, that the dysgraphia is a result of dyspraxic performance in any of the disconnection syndromes. The dyspraxia and the dysgraphia may result independently from the disconnection of hand use from language.

*Pure dysgraphia* is a disturbance of writing unexplained by a more primary deficit in motor function, in language, in praxis or in spatial manipulations. There is a long history of reports which attempt to confirm the existence of pure dysgraphia and thus to localize the area of brain dysfunction responsible for pure dysgraphia. Exner (1881) cited evidence in favor of a motor graphic center in the posterior middle frontal gyrus. Henschen (1922) added a possible sensory graphic center in the angular gyrus. It is clear that most lesions in either of those locations will produce broader disturbances in language and praxis. Cases of pure dysgraphia are rare, and there is less consensus about the nature of pure dysgraphia than about any other graphic disorder.

A handful of reports of pure dysgraphia have appeared in recent years

(Hecaen and de Ajuriaguerra, 1963; Dubois *et al.*, 1969; Assal *et al.*, 1970; Aimard *et al.*, 1975; Rosati and DiBastiani, 1979). These reports confirm that relatively isolated disturbances in writing can occur, but they shed less light on a specific mechanism or localization. Most suggest that the problem is in the left frontal lobe. In these cases the writing disorder consisted of 1) normal letter formation; 2) normal copying; 3) written spelling errors which were mostly substitutions.

Russell and Espir (1961) found examples of pure dysgraphia with gunshot wounds of the deep left posterior parasagittal parietal region. The superior posterior parietal region was implicated also in the case of Kinsbourne and Rosenfield (1974). The dysgraphia in their patient was not strictly pure (mild anomia, Gerstmann's syndrome), but the striking deficits in written spelling compared to oral spelling suggested a specific disorder in the transformation of letter sounds or names into a visual representation for writing. Basso *et al.* (1978), in a retrospective analysis of 500 dysphasics, found 2 with pure dysgraphia. These patients had no dyspraxia and no dysphasia. Writing errors were prominent and involved linguistic elements (neologisms and misspellings). Both patients had lesions in the left superior, posterior parietal region.

Although the available literature does not allow much speculation about mechanisms of pure dysgraphia, there is a plausible anatomical basis: a superior parietal region, intimately tied to the cross modal associations of the angular gyrus, but relatively more involved in the visual-kinesthetic-constructional aspects of phoneme-grapheme conversions. Dysgraphia on a parietal basis is unusual because the responsible superior parietal region lies outside the usual area of involvement in cerebrovascular disease. In addition, a small increase in lesion size would produce additional deficits such as dyslexia.

Chedru and Geschwind (1972) raised several objections to all of the cases of pure dysgraphia up to that time. They asserted that the patients reported had not been suitable for brain-behavior correlations. There is a high percentage of tumors; many cases had increased intracranial pressure, diminished attention, and frank confusion, or were examined too soon after onset of their disease. Some of the cases were simply examples of callosal dysgraphia (see above). In addition, the writing abnormalities of the reported cases were similar to the abnormalities seen in patients with acute confusional states. Most of the literature on pure dysgraphia is susceptible to these criticisms, but the particular disorder in written spelling in the absence of dysphasia or dyspraxia (Basso *et al.*, 1978) may be an important exception.

# Special Clinical Forms of Dysphasia   F

Syndromes of dysphasia may vary from the forms described in the preceding sections, or certain aspects of dysphasic syndromes may appear more prominently than others, depending on associated clinical conditions, language background of the patient, history of left-handedness, lesion localization, or individual differences. In this section we consider clinical features of some of these special forms of dysphasia.

## 1. Crossed Dysphasia

When the term "crossed dysphasia" was introduced in 1899 by Bramwell, it referred to either of two conditions: right hemiplegia and dysphasia in a left hander, or left hemiplegia and dysphasia in a right hander. With the passage of time the term has taken on a single meaning. Today, "crossed dysphasia" refers specifically to dysphasic syndromes found in right-handers with damage to the right hemisphere.

The syndrome of crossed dysphasia is rare, with estimated incidence ranging from 0.4 % (Hecaen *et al.*, 1971) to about 2 % (Zangwill, 1967). "Negative" proof of the existence of the syndrome has also been presented. For example, Boller (1973) detailed the observation of a right-handed man with total destruction of the posterior third of the superior temporal gyrus (Wernicke's area) in the left hemisphere who had no evident language deficit. Language, in this patient, may have been organized in the right hemisphere.

When damage occurs to the right hemisphere of a right-handed person resulting in a dysphasic syndrome, one must assume that language was structurally (anatomically) based in the right hemisphere of that individual, or at the least, that language was not totally dependent on left hemispheric mechanisms. Studies of anatomical asymmetries of the human brain indicate that "right-sided cerebral dominance for speech may be more common than is generally thought" (Galaburda, LeMay, Kemper, Geschwind, 1978).

Zangwill (1967) and Hecaen (1977) have suggested that the clinical patterns of dysphasia found in right-handers with right hemispheric lesions do not resemble the "classical" clinical forms of dysphasia produced by left hemispheric lesions in right-handed persons. Zangwill comments that there may be something "special" about such patients. Our own experience suggests that the clinical features of crossed dysphasia resemble those of dysphasic syndromes in left handers (Hecaen and Albert, 1978). In their descriptions of crossed dysphasia most authors point out that agraphia and agrammatism seem to occur most frequently, and appear regardless of lesion localization (Ettlinger *et al.*, 1955; Angelergues *et al.*, 1962; Barraquer-Bordas *et al.*, 1963; Clarke and Zangwill, 1965; Brown and Wilson, 1973). Comprehension of spoken language is generally preserved, and naming is only mildly impaired. Emphasis has been laid on the commonly associated non-linguistic findings, such as confusion, memory disorders, attentional defects, personality changes and perseveration (Marinesco *et al.*, 1932; Stone, 1934).

As for etiology of crossed dysphasia, a critical review conducted by Boller (1973) indicated that acute vascular lesions account for only 23 % of the cases, with the majority being caused by tumor or trauma. These figures are in sharp contrast with those for dysphasia in the general population, for which vascular accidents are the most common cause. Perhaps some of the "special" features of crossed dysphasia are caused by bilateral effects of head trauma or bilateral pressure effects of an intracerebral tumor.

## 2. Dysphasia in Polyglots

Clinical experience with dysphasia in polyglots has led most specialists to agree on a set of observations and assumptions: 1) in the vast majority of cases the patterns of dysphasia in each language of a polyglot resemble the classical clinical patterns of dysphasia in monolinguals; 2) if different patterns of dysphasia appear in the different languages of a polyglot, the differences can be accounted for by different premorbid degrees of fluency in each language; 3) the syndromes of dysphasia in each language of a polyglot tend to parallel each other qualitatively (even if not quantitatively) following onset of the illness and during recovery; 4) if the polyglot is right-handed, the lesion producing the dysphasia is usually in the left hemisphere; it is most unusual for right hemispheric lesions to produce dysphasia, even in polyglots.

Despite general agreement on the set of observations listed above, specialists continue to be intrigued by the problem of dysphasia in polyglots for several reasons: a small minority of polyglot dysphasics have syndromes which do not conform to the rules. In particular, qualitatively different patterns of dysphasia have been observed in individual polyglots at the same time (Bychowski, 1919; Albert and Obler, 1975; Silverberg and Gordon, 1978); and differential (non-parallel) recovery patterns have been described for each of the languages affected by the dysphasic deficit (see Paradis, 1977, or Albert and Obler, 1978, for detailed reviews of these cases). In terms of incidence, these variant cases must be exceedingly rare; of the tens of thousands of cases of dysphasia in polyglots which must have occurred in the past 150 years, only about 135 cases of atypical dysphasia (*i.e.* with different syndromes or differential recovery patterns) have been reported in all of the available world's literature in any language (Paradis, 1978, personal communication).

Even though the number of variant cases is small, these cases have attracted wide attention and abundant theorizing. Theories to account for differential patterns of recovery must explain those clinical situations in which an individual is dysphasic in one language but not in another (*e.g.* Halpern, 1941); more severely dysphasic in the primary language than in a second language (*e.g.* Minkowski, 1963); more severely dysphasic in the most recently and most frequently used language than in a language acquired in childhood (*e.g.* Dreifuss, 1961); or in which an individual manifests at the same time and from a single lesion a non-fluent dysphasia in one language and a fluent dysphasia in another (*e.g.* Silverberg and Gordon, 1978).

An early theory to account for differential recovery was based on Ribot's (1881) general theory of memory disorders which stated that earlier learned items are better preserved in brain damage, and that, in recovery from memory loss, earlier learned items return before items learned later in life. This theory, applied to dysphasia in polyglots, implied that in cases of differential recovery, the language learned first recovers first. In 1895 Pitres proposed another rule: the polyglot patient first regains comprehension of his most recently used language; next he recovers ability to speak this language; then he regains the capacity first to understand and then to speak the other language. Krapf (1957) and Minkowski (1965) hypothesized that affective factors help determine which language would return first.

Some authors have suggested that differential recovery may result from different patterns of anatomical organization of the languages (Scoresby-Jackson, 1867; Gloning and Gloning, 1965; Albert and Obler, 1975). Others (*e.g.* Charlton, 1964; L'Hermitte *et al.*, 1966) have denied this position, citing as evidence studies of parallel recovery in *unselected* groups of polyglot dysphasics, and concluding that the majority lose and recover their languages in proportion to the premorbid degree of fluency in the language. Following a thoughtful and thorough analysis of more than 100 cases of polyglot dysphasia, Paradis (1977) concluded that several factors may contribute in varying degrees to recovery patterns, including order in which the two languages are learned, degree of proficiency in each language, affective attitudes towards each language, site and size of lesion, and the general biological condition of the patient.

In a recent monograph on the neurological basis of bilingualism, Albert and Obler (1978) reviewed neurolinguistic, neuropsychological, and neurological studies of bilingualism and concluded that language is organized in the brain of a bilingual in a manner different from that which might have been predicted by studies of cerebral organization for language in monolinguals. Studies of bilinguals demonstrated a major right hemispheric contribution to language in addition to the left hemispheric contribution. This conclusion remains to be confirmed by studies in other laboratories.

## 3. Dysphasia in Left-Handers

Two issues have been of principal concern to specialists studying break-downs of language function in left-handers: the question of cerebral dominance for language, and the clinical characteristics of the dysphasic syndromes. For both issues the situation is different with left-handers as compared with right-handers.

It has been stated that more than 99 % of right-handed people have dominance for language in the left hemisphere (e.g. Benson and Geschwind, 1977). The picture is not so clear for left-handers. Estimates of the degree to which the left or right hemisphere is dominant for language in left-handers vary considerably, and the evidence to support these estimates varies, as well, from uncontrolled, clinical, often anecdotal reports (e.g. Broca's (1865) claim that the right hemisphere is dominant for language in all left-handers) through carefully detailed, systematic studies of dysphasia in left-handed subjects with unilateral cerebral lesions. These have suggested that the left hemisphere is dominant for language in more than 50 % of all left-handers (e.g. Goodglass and Quadfasel, 1954; Hecaen and de Ajuriaguerra, 1963; Subirana, 1969; Luria, 1970).

A coherent summary is that provided by Goodglass and Quadfasel (1954): no direct and necessary correlation can be established between handedness and cerebral lateralization for language; left hemispheric dominance for language is more frequent than right-handedness, and right hemispheric dominance for language is much less frequent than left-handedness. A review of several series indicates that 20—30 % of left-handers with dysphasia have right hemispheric lesions. To these observations we should add those of Gloning et al. (1969). These authors studied a large series of left- and right-handed subjects with brain damage matched for site of lesion. They found that for left-handers a lesion in either hemisphere was as likely to produce dysphasia as a left-sided lesion in a right-handed patient. They concluded that left-handers may have language organized in both hemispheres.

Studies of anatomical asymmetries in the human brain may provide a structural basis for these behavioral observations. A number of anatomical asymmetries have been demonstrated in the brains of right-handers but not in left-handers in regions thought to be related to language function. For example, the left occipital lobe is wider than the right (LeMay, 1976); the sylvian fissure is longer on the left (LeMay and Culebras, 1972). From these and other observations LeMay and Geschwind (1975) discerned a general pattern: 1) that brains without a particular asymmetry are more common in the left-handed; 2) that left-handers are more likely than right-handers to show an asymmetry in the opposite direction; 3) even in those left-handers who have an asymmetry in the same direction as right-handers, the asymmetry is less marked.

If the brains of left- and right-handers are different anatomically, this fact might account for the clinical and experimental observations that lesions in the same regions of the brain produce different patterns of behavioral deficit in left- and right-handers. Lesions in the left hemisphere which produce classical forms of dysphasia in right-handers, do not always do so in left-handers. Typically, dysphasia in a left-hander is milder than in right-handers, regardless of the hemisphere damaged (Luria, 1970; Hecaen and Albert, 1978). In addition, left-handers generally recover more quickly and more thoroughly from dysphasia than do right-handers. The "left hemispheric syndrome" in left-handers corresponds in many respects to that in right-handers; however, disorders of auditory comprehension, of writing and of spelling are significantly less frequent in left-handers than in right-handers, while dyslexia and spatial dyslexia are more frequent (Hecaen and Sanguet, 1971).

# 4. Dysphasia in Deaf-Mutes

One major question that has been asked about dysphasia in deaf-mutes is this: To what extent is the neurological basis of communication skills in the deaf-mute similar to that in persons with normal hearing; are the biological (neurological) systems which underlie communication in the deaf-mute the same as or different from those which underlie language in the person with normal hearing? If the communication skills of the deaf-mute use anatomical structures and physiological systems which are independent of those which form the basis for language in the person with normal hearing, perhaps these communication skills could be taught to the person with normal hearing who becomes dysphasic as a result of brain damage in the "zone of language" (Dejerine, 1914).

Studies of dysphasia in deaf-mutes could potentially provide answers to this question, by demonstrating dissociations of behavioral deficits (sparing of some communication skills, loss of others) in association with discrete cerebral lesions. Available data are insufficient, however, to provide any meaningful response. The reported cases of dysphasia in deaf-mutes have been too few and too dissimilar (in terms of age of onset of deafness, location of lesion, means of communication used by the deaf-mute) to allow systematic analysis.

Consequently, what we shall do in this section is summarize on a case by case basis reports from the literature on dysphasia in deaf-mutes, and comment on key points raised by this collection of individual cases. We might add that, although we shall continue to use the term "deaf-mute" because it is the term more commonly used in the literature, we agree with Sarno, Swisher, and Sarno (1969) that the term "congenitally deaf" is more accurate, since many who are born without hearing learn some speech.

In 1896 Grasset described a case of "dysphasia in the right hand of a deaf-mute". This patient generally communicated with his right hand. Following a vascular lesion to the left hemisphere, the patient was no longer able to communicate, by sign language or by writing, with his right hand, but was able to communicate with his left hand. Comprehension of sign language remained intact.

Critchley (1938) reported on a case of dysphasia in a partial deaf-mute. This patient had been able to use either hand for communication (with sign language or by writing), could read lips, and could mouth and even speak words. Following an acute vascular accident with right hemiplegia, he lost all ability to express himself by any means, and seems, also, to have lost the ability to read lips. After a period of modest recovery, oral expression remained severely dysarthric and agrammatic, and manual communication remained even more impaired.

Leischner (1943) provided one of the few available cases of dysphasia in a deaf-mute with anatomical (autopsy-proven) evidence of lesion-localization. The patient was a 64-year old deaf-mute polyglot who suffered a thrombotic stroke involving the left supramarginal gyrus. The dysphasic syndrome affected both expression and comprehension. Spoken and manual language were markedly impaired, but during the recovery phase manual language recovered more than spoken language. Comprehension of sign language was moderately impaired, although comprehension of written language was relatively spared. Following the initial post-stroke period, the patient was able to express himself by writing, although he tended to mix his two languages. Leischner concluded that his patient had a general disturbance in ability to use any form of symbolic expression.

The case described by Tureen, Smolik, and Tritt (1951) is that of a 43-year old congenitally deaf-mute man whose presenting condition was one of acute confusional state with right hemiplegia. When he recovered from the confusional state he was found to have lost what little ability he had previously had to speak, but that he had retained the ability to read written words and to read lips. By contrast, his ability to communicate manually —either to express himself or to understand signs—was markedly impaired. He was found to have a left hemispheric tumor subcortically located below Broca's area. Following removal of the tumor he recovered the capacity to say a few words and to communicate by sign language (both reading and expressing himself). No dyspraxia was observed throughout the course of the illness.

Douglas and Richardson (1959) described a congenitally deaf-mute woman who suffered a left hemispheric stroke as a result of which she had a right hemiplegia, with acalculia and astereognosia. Using finger spelling, she was able to recognize single letters, but not words. She was also unable to understand conventional descriptive (symbolic) gestures. Her ability to express herself manually was severely impaired, both for spontaneous expression and on imitation. Perseverations intruded in her behavior. Errors resembled those seen in a person with normal hearing who develops an anterior dysphasia.

Sarno, Swisher, and Sarno (1969) presented a detailed and comprehensive description of a 69-year old right-handed man born with a profound hearing loss who sustained a left hemispheric stroke with right hemiplegia and dysphasia. This patient had used basic sign language, finger spelling, writing, reading, and lip reading before his stroke. He also *spoke*, in short phrases, frequently without phonation but with much lip articulation. His dysphasic syndrome involved all modalities of expressive communication. The authors stressed that, apart from the difficulties of expression for symbolic communication, their patient had no dyspraxia—neither for bucco-facial nor for limb movements. Comprehension was also impaired initially; but recovery of comprehension skills was better than that of expressive skills. These authors were struck by the many ways in which the deficits manifested by their patient resembled those of a classical motor dysphasia in a person with normal hearing. They concluded that dysphasia "in the congenitally deaf is entirely equivalent to that in normal hearing people" and that "the areas and processes in the dominant hemisphere which subserve communication functions are the same for the congenitally deaf as for the hearing person".

Our own position on this issue, while tentative due to the small number of cases available for study, is somewhat different from that of Sarno *et al.* (1969). While we agree that the same neurological structures may underlie certain aspects of communicative abilities in deaf-mutes and in persons with normal hearing, it seems to us that the cognitive strategies involved in manual speaking versus auditory-oral communication are sufficiently different that different physiological systems may be involved. When differential patterns of dysphasia appear in the manual communication system as opposed to the spoken language system of the deaf-mute, these patterns may be the result of disruption of different sets of cognitive processes which control the different communication systems available to the deaf-mute. When the same pattern of dysphasia appears in all available communication systems of the deaf-mute, the lesion may simply have been large enough to affect all anatomical structures necessary for each different communication system.

# 5. Dysphasia in Dementia

To study fully the issue of dysphasia in dementia one should first consider the nature and relationship of normal language and intelligence, and then consider independently the breakdown of each of these behavioral capacities following brain damage. Such an approach would carry us well beyond the intended scope of this book, which is designed to focus on clinical issues. For the purposes of this book we shall consider only a limited portion of the topic suggested by the title of this section. First we shall review findings of those who have studied language in a variety of dementing syndromes; then we shall describe patterns of dysphasia which occur in two broad categories of dementia: the dementias caused by cerebrovascular disease (multi-infarct dementia), and the dementias of the Alzheimer's Disease-Senile Dementia complex.

*Language in Dementia.* Major studies of language in dementia include those of Seglas (1892), Stengel (1964), Critchley (1964) and Irigaray (1967, 1973). A general conclusion may be drawn from these studies that disorders of linguistic performance in dementia may be distinguished from specific dysphasic syndromes. Another general conclusion is that although there may be qualitatively different types of dementia, and the language defects may vary with the type of dementia, nonetheless there are certain characteristics of impaired language performance found in most types of dementia.

These characteristics include the following: 1) breakdown in logical associations of spoken discourse, resulting in incoherence of output; 2) reduction of lexical stock, manifested by naming deficit; 3) simplification of syntax; 4) perseveration; 5) echolalia; 6) introduction of improbable or unlikely phrases; 7) tangentiality; 8) tendency for the above deficits to increase as length of conversation increases.

From our own studies of language in dementia (Silverberg and Albert, 1975; Obler and Albert, unpublished data) some general observations may be extracted. Language disorders seen in patients with senile dementia, but without significant dysphasia, are heavily influenced by a combination of non-linguistic factors. These factors include impairment of memory, reduced ability to deal with normally paced speech (*i.e.* a problem of rate of information processing), perseveration, and reduced attention span.

*Dysphasia in Multi-Infarct Dementia.* In patients with dementia due to cerebral arteriosclerosis (which represents 10—15 % of the demented population, according to Terry (1976)), infarction in language areas of the left hemisphere may result in a dysphasic syndrome which is modified by non-linguistic deficits as described above in the section on "language in dementia". Three types of onset are typical. Dysphasia may appear suddenly in a previously demented patient; or, dysphasia may precede the general intellectual impairment, which develops slowly in a step-wise fashion; or, against a background of intellectual deterioration, minor pathological vascular events may occur repeatedly in the left hemisphere, producing a slowly evolving dysphasic picture.

The commonest type of dysphasia to occur under these conditions is anomic dysphasia; the next most common type is one or another variety of Wernicke's dysphasia. The language disorder may appear at first to be more severe than it actually is, since the patient will often perform better after a period of personal human contact and stimulation. Even under the best, warmest, and most supportive of examination settings, however, the patient's stock of ideas seems to be reduced. Features of thought disorder may mix with paraphasias and verbal amnesia to produce a pattern of total jargon dysphasia. Defects of writing and of auditory comprehension are often marked. Perseveration and echolalia are characteristic features.

*Dysphasia in the Alzheimer's Disease-Senile Dementia Complex.* The Alzheimer's Disease-Senile Dementia (AD-SD) complex represents the majority of dementing syndromes. In their extensive study Sjogren *et al.* (1952) found some form of dysphasia in every one of their cases of Alzheimer's disease. Review of early- and late-onset dementias of the AD-SD type suggest the following order for dysphasic syndromes in terms of frequency of occurrence with the dementing syndrome: anomic dysphasia is most common, occurring in more than 95 % of AD-SD; posterior (Wernicke's type) dysphasia is next, occurring in more than 80 %; agraphia is frequent—about 80 %; and alexia is frequent—about 75 % (Sjogren *et al.*, 1952; Delay and Brion, 1962). The primary clinical features are word-finding difficulties and inability to understand spoken or written language, together with perseveration and echolalia. For reasons as yet unknown, Broca's dysphasia is uncommon in the Alzheimer's Disease-Senile Dementia complex.

# Part III
# Therapy of Dysphasia in Adults

Although descriptions of dysphasia date from the time of Hippocrates (Benton and Joynt, 1960) there was little reference to dysphasia rehabilitation until 1880. In that year Charles Mills, an American physician, cited a case in which a patient made remarkable advances through "persistent efforts to re-educate his stricken brain to regain powers of speaking and writing". Mills concluded that although it was fashionable to look upon cases of dysphasia as absolutely hopeless, much could be done for them.

Nearly one hundred years later the medical community continues to challenge Mills' statement, and to question the efficacy of dysphasia rehabilitation (Benson, 1979). Speech pathologists and physicians alike ask whether dysphasia treatment is effective, and if so, what approaches have proven viable. The answers to these questions may lie within the extant literature on dysphasia rehabilitation. Since much of this literature has already been reviewed in excellent detail (Darley, 1970; 1975; Eisenson, 1975; Sarno, 1974), we will cite the past only in so far as it is felt to affect the present, and will examine the present only in so far as it provides possible direction for the future.

9*

# A   Is Dysphasia Rehabilitation Effective?

## 1. Introduction

Aphasiologists are no longer satisfied with impressionistic judgements as to the effectiveness of dysphasia rehabilitation. Instead they seek scientific evidence, through formal studies, that treatment has a significantly positive effect on dysphasic patients. When undertaking any treatment study, certain variables which may confound the results must be identified and controlled. The most potent of the variables to be considered in dysphasia rehabilitation is the phenomenon known as *spontaneous recovery*. This is the period during which structurally undamaged portions of the brain regain function following insult. If treatment is administered during the period of natural recovery, then the investigator has the task of differentiating effect of treatment from effect of spontaneous improvement. There is, however, little agreement as to exactly how long spontaneous recovery takes place. Vignolo (1964) retrospectively examined the evolution of dysphasia in 69 non-traumatic patients and found that patients seen after 2 months had only a limited possibility of spontaneous recovery compared to patients seen before 2 months. Culton (1969) studied the recovery patterns of 21 dysphasics and concluded that the greatest degree of recovery takes place in the first month post onset, and that little more occurs after 2 months. Butfield and Zangwill (1946) and Luria (1963), on the other hand, state that spontaneous recovery takes place for as long as six months, particularly in traumatic dysphasias. Other authors use a 3 month cut off point (Sarno and Levita, 1971). Despite the disagreement as to the length of the spontaneous recovery period, the phenomenon is real, and must be reckoned with in any dysphasia treatment study.

A second confounding variable has to do with *etiology*. Butfield and Zangwill (1946) found that dysphasias caused by trauma resolve better than dysphasia caused by strokes. Even within the stroke group differences exist.

Johnson (1975), for example, found that patients with hemorrhages have a different recovery rate than patients with thromboembolic disease.

A third confounding variable in any dysphasia treatment study is that different dysphasic *syndromes* recover differently. Kertesz and McCabe (1979) found, for example, that anomic patients make the best recovery, and patients with Wernicke's dysphasia who do not initially present with jargon dysphasia recover better than those who do. This study also confirmed the often reported finding that most globally dysphasic patients make very limited recovery. Earlier Vignolo (1964) hypothesized that anarthria may be responsible for limiting recovery in expressive skills.

In addition to the variables mentioned above there is some evidence that *age* may be an important factor in recovery from dysphasia. In Vignolo's (1964) retrospective study of dysphasia improvement appeared strongly to favor younger patients. Sands, Sarno, Shankweiler (1969) studied the progress made by 30 patients receiving non-specified speech therapy, and found that age was apparently the most potent variable. Smith's (1971) data also suggest that recovery from dysphasia may be age-related, with elderly patients showing diminished recovery. Conversely, Culton (1969) found that age was not a factor in early spontaneous recovery and that his older patients showed a trend toward greater recovery. Finally, Sarno (1979) reported in a recent study which carefully examined the specific influence of age on recovery patterns that no patient should be denied dysphasia rehabilitation solely on the basis of advanced age. Helm (1978) found that age was a non-significant variable in predicting response to a specific treatment of melodic intonation therapy.

*Educational level* appears to be a non-significant variable in recovery from dysphasia, according to Sarno, Silverman, and Levita (1970), Sarno, Silverman, and Sands (1970), Smith (1971) and Helm (1978).

In addition to the variables discussed above, there are several behavioral factors which are felt to have significant effects on the course of recovery from dysphasia. Eisenson (1949) felt that patients who have outgoing personalities, and modest levels of aspiration, have a better prognosis than patients who are euphoric, rigid, or dependent. Benson (1973) discussed several psychological states which may adversely affect response to dysphasia rehabilitation. Among these are depression, paranoia, and euphoria. These variables, of course, are difficult to measure, but since they may have a significant effect on treatment response they should be considered in studies of dysphasia rehabilitation.

## 2. Measuring Response to Treatment

There are various ways to measure the dysphasic patient's response to treatment, some more refined than others. The least refined method is to rate communication skills according to a scale. Of course, the more points on the scale, the more refined the measurement, but scales are essentially impressionistic and subjective. Unfortunately, several of the major dysphasia treatment studies have used three point scales to rate the patient's language skills as either much improved, improved, unchanged (Butfield and Zangwill, 1946), good, fair, unchanged (Godfrey and Douglass, 1959), or recovered, improved, unchanged (Vignolo, 1964). A more refined scale was used by Sarno, Silverman, and Sands (1970) to measure changes in language performance in a treatment study of global dysphasia patients. These latter investigators employed the *Functional Communication Profile* (Taylor, 1965), which rates five categories of communication along an eight point continuum. Although Basso, Capitani, and Vignolo (1979) used a 5 point scale to measure changes in language skills, the scale was based on raw scores earned on a variety of formal tests.

Another objective method of measuring response to dysphasia treatment is to compare pre- and post-therapy scores obtained on standardized dysphasia tests. For example, the *Boston Diagnostic Aphasia Examination* (Goodglass and Kaplan, 1972) was used to study the effects of a specific treatment (melodic intonation therapy) on groups of severely dysphasic patients (Sparks, Helm, Albert, 1974; Helm, 1978), and the *Porch Index of Communicative Ability* (Porch, 1971) was used to compare the effects of group versus individual treatment in the Veterans Administration Cooperative Study on Dysphasia (Wertz, 1978).

Thus we find that the tool which is used to measure response to dysphasia treatment is an important consideration when either planning a study, or reviewing the conclusions drawn by other investigators.

# 3. Studies of Dysphasia Treatment

There are two major approaches to studying the effects of dysphasia rehabilitation: the group study approach, and the case study approach. If one simply wishes to compare the general effects of treatment versus no treatment on numbers of patients, then two groups matched for such factors as age, etiology, and time past onset are required. It is, however, possible to compare the effects of treatment and natural recovery on a single individual. If one is interested in studying the effects of a specific treatment on numbers of patients then at least three matched groups, one of whom receives the specific treatment, one of whom receives an alternative treatment, and one of whom receives no treatment, are required. Again, it is possible to study individual dysphasics, but instead of merely investigating treatment versus no treatment, one can compare various specific approaches.

*Treatment versus No Treatment: The Group Study Approach*

Most rehabilitation centers have neither the patient nor financial resources to run proper group studies using untreated controls. Even when these resources are available, ethical considerations may prohibit investigators from denying a group of patients an opportunity to receive treatment. Sometimes, however, natural conditions allow for this design. One informal example of a naturally occurring treatment versus no treatment study is found in Eisenson (1947). Due to a World War II bureaucratic error, a large group of dysphasic veterans was not transferred to the military hospital's dysphasia unit. When these patients were "discovered" they seemed not to have made gains equal to the patients who had been treated for dysphasia. In addition, many of the untreated group were felt to be discouraged and to have "psychological problems".

A more scientific study of treatment versus no treatment was undertaken by Smith in 1972. Pre and post treatment language test scores of 80 treated and 15 untreated dysphasics were contrasted. The findings indicate that treated patients improve in language performance beyond the level which might be expected with spontaneous recovery alone.

Two more recent studies which compared treated and untreated groups have been published by Italian investigators. The first is described by Basso, Faglioni, Vignolo (1975). For primarily geographical reasons, a group of 94 patients was unable to travel to treatment centers. These patients did not recover to the same extent on a 4 point scale as the 94 patients who had treatment. One might raise questions as to the degree of similarity between the treated and untreated groups, as the over-all quality of life may differ significantly between rural and the more urban environments close to treatment centers. The findings, however, do suggest that dysphasic patients who receive treatment improve to a greater extent than those who are left untreated.

Basso, Capitani, and Vignolo (1979) studied the influence of treatment on 162 patients who were compared to 119 untreated controls. They found that formal language rehabilitation has a positive effect on speaking, listening, writing and reading skills, if administered for at least six months for no less than 3 sessions per week.

*Treatment versus No Treatment: Single-Subject-Time-Series Approach*

An alternative to large treatment studies has been proposed by La Pointe (1977). He recommends the use of a time series design in studying the effects of treatment. This design examines the effects of a treatment variable on behavior as measured by the consecutive presentation, removal and representation of that variable. Davis and Bisset (1977) used this single subject research design in evaluating dysphasia rehabilitation. These investigators measured the effects of treatment on 5 separate patients by looking for a "stair-step" effect which corresponds to alternating periods of treatment, no treatment, treatment. Davis and Bisset feel that their data generally indicate that intensive treatment does have a decisive influence on the course of recovery from dysphasia. Specifically, it was found that some patients, especially Broca's dysphasics, made substantially more improvement with treatment than could be expected spontaneously in the first six months following onset.

Although both La Pointe (1977) and Davis and Bisset (1977) use the Porch Index of Communicative Ability (Porch, 1971) to measure the "stair-step" effect of single-subject treatment, any test can be used. For example, the following results were obtained over time with the Boston Diagnostic Aphasia Examination (Goodglass and Kaplan, 1972). Note that the largest gains in confrontation naming scores were achieved during specific treatment and that these gains were maintained during the untreated periods. These results dramatically illustrate that treatment can affect the performance of dysphasic patients, and that this effect is not transient.

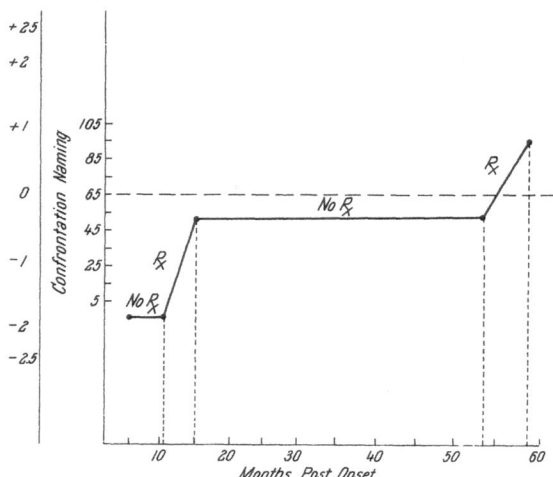

Fig. 9. Single subject research design for evaluating the effects of treatment

The case study approach to evaluating the effects of treatment is not new. In 1935, Weisenberg and McBride did a controlled study of the effects of re-education on one patient. The first of a series of formal examinations took place two months post onset. For the next five months the patient was untreated following which the tests were re-administered. The patient was treated then for six months for approximately two hours a week plus home assignments. The tests were re-administered for a third time and the results showed that far more improvement had taken place during the final six months during which treatment occurred than during the first nine months without treatment. In reviewing this and similar cases, Weisenberg and McBride concluded that training increases the amount of recovery, improves performance in specific areas such as articulation, and possibly prolongs the improvement period. As in most case study presentations these authors carefully described the nature of the treatment, as well as the nature of the dysphasia.

## Comparing Specific Treatments and No Treatment: Group Study Approach

One of the few attempts to investigate the effects of specific treatments was undertaken by Sarno, Silverman and Sands (1972). These investigators compared the effects of programmed instruction, traditional instruction, and no instruction on three groups of patients who were matched for etiology (cerebrovascular accident), time post onset (at least three months), and type of dysphasia (global). Since no differences were found between groups on pre- and post-Functional Communication Profile Scores the authors concluded that globally dysphasic patients do not benefit from speech therapy.

## Comparing Specific Treatments and No Treatment: Case Study Approach

Although there is to date no report of the use of the single subject design to compare the effects of various specific treatments and no treatment on a single individual, such an approach is possible. The patient would be exposed to alternating periods of no treatment, treatment A, and treatment B. If the patient consistently showed significantly greater improvement with a particular treatment, then that treatment may be considered the most effective of the three.

## Using Patients as Their Own Controls: An Alternate Group Treatment Design

One way to circumvent the ethical problem of including untreated controls in a group treatment study is to use patients as their own controls. While this is a less rigorous approach to studying treatment effect, it is not without merit. Such an approach was taken in the melodic intonation therapy study (Sparks, Helm, Albert, 1974; Helm, 1978). This study sought to determine the effects of a special treatment on the speech output of severe verbally impaired dysphasic patients. The patients served as their own controls in the following manner. 1. Only patients who had remained severely impaired in verbal expression for at least four months were admitted to the study. By setting the lower limit for eligibility at four months it was hoped that significant post-treatment gains in speech could be attributed to the effects of treatment and not to spontaneous recovery. 2. Only patients who had received a previous course of language rehabilitation but had remained severely impaired in verbal expression were admitted to the study. It was felt that if a patient did not respond to a different course of treatment which was administered during the optimal recovery period, but did respond to a subsequent new form of therapy then one might assume that the new method had special merit. Analysis of pre- and post-treatment Boston Diagnostic Aphasia Examination (Goodglass and Kaplan, 1972) naming scores obtained on twenty patients revealed that melodic intonation therapy is a particularly effective method with patients displaying the main characteristics of the syndrome known as severe phonemic articulatory disorder of speech (De Renzi, Pieczuro, and Vignolo, 1966) with sparse effortful, dysprosodic output which may be limited to a quick succession of a few recurrent, meaningless phonemes. The following are transcripts of the pre- and post-melodic intonation therapy conversational speech of one such patient. The first sample was obtained immediately following four months of traditional treatment and immediately preceding initiation of the intoned treatment.

Pre-Melodic Intonation Therapy Conversation

Examiner: "How are you today?"
Patient: "Nee nay—nee, no, no."
Examiner: "Have you ever been in this hospital before?"
Patient: "Ah—no, nee, nay, no, no nee."
Examiner: "Do you think we can help you here?"
Patient: "Nee, nee, nay no, nah, nee."

The second sample was obtained immediately following a three month course of Melodic Intonation Therapy.

Post-Melodic Intonation Therapy Conversation

Examiner: "How are you today?"
Patient: "Oh—pretty good."
Examiner: "Have I ever tested you before?"
Patient: "I don't know."
Examiner: "I did when you first came in. Maybe you don't remember. Do you know why I'm testing you now?"
Patient: "Dis week. I'm going to go home."
Examiner: "That's right. Do you know when?"
Patient: "One, two, three days."

Since this patient and others like her did not respond to the more traditional approaches to language rehabilitation which were administered during the optimal recovery period, but later showed significant response to the new method, a strong case can be made in favor of a course of melodic intonation therapy for this particular segment of the dysphasic population.

Basso, Capitani, and Vignolo (1979) describe three patients who received treatment after the period of spontaneous recovery (11 years; 2 years 2 months; 7.5 months). Each of the patients improved dramatically with treatment.

## 4. Conclusion

In attempting to answer the question concerning the efficacy of dysphasia rehabilitation, we find good evidence in both group and case studies that nonspecific dysphasia rehabilitation is *generally* effective and preferable to leaving the patient untreated. There is further evidence that certain types of dysphasics will make significantly more improvement when treated in specific ways than could be expected if no such treatment were undertaken. This leads to the second of the two questions posed in the introduction, which has to do with specific approaches to dysphasia rehabilitation.

# What Approaches to Dysphasia Rehabilitation Are Felt to Be Most Effective? **B**

## 1. Introduction

Speech pathologists are called upon to rehabilitate adults with disorders which range from global dysphasia with severe impairment in all language modalities, to anomic dysphasia with impairment only in substantive word finding. There is little reason to expect that disorders which vary widely in nature should or can be treated in a similar manner. There are, of course, some general principles which can be applied to any rehabilitative process and dysphasia is no exception. Such clinicians as Backus (1937) and Schuell, Jenkins and Jiméniz-Pabón (1964) provide us with general principles of treatment, and these principles are no less appropriate today than when they were written. Few of us would dispute, for example, that speech processes operate with greater facility when the individual experiences a reasonable degree of social adequacy (Backus, 1937), or that the clinician should elicit and not force the response (Schuell, Jenkins and Jiménez-Pabón, 1964). But while such principles may guide us in the treatment process, we must have a specific method in mind when sitting across from the patient.

Martin (1977) pointed out that all treatment should proceed from a theoretical view as to the nature of the disorder. This allows us to determine a rationale for treatment. The rationale in turn leads to specific techniques and goals. He illustrated this process in the following manner. If we view dysphasia as a loss of language, then we have a rationale for replacing the lost information through teaching techniques. If we view dysphasia not as a loss of language but as a reduction of efficiency in gaining access to linguistic knowledge, then we have a rationale for stimulating the patient to improve his accessing or retrieval strategies.

These two views of dysphasia seem to have determined the approaches to rehabilitation since Charles Mills wrote the first treatment paper in 1904. Mills viewed dysphasia as a loss of language and felt that dysphasics must be retaught as children. Lennenberg (1967), on the other hand, stated that in contrast to the normal small child, adult dysphasics do not learn language, because their problem is not that they do not know language, but rather can no longer make use of language that they have already learned. Schuell, Jenkins and Jiménez-Pabón (1964) also made it clear that they considered the damaged adult brain to be different from the child's developing brain. In their view, the clinician does not teach but instead tries to communicate with the patient and to stimulate the disrupted language processes to function maximally. Wepman (1953) also spoke of stimulation. He felt that the role of the clinician was to provide the patient with stimulative material at the time when the central nervous system was most capable of using it for facilitation of the cortical integration which leads to language performance.

And so it is that throughout the history of dysphasia rehabilitation, and continuing to the present, two opposing views of dysphasia—either the loss of language or the impaired capacity to gain access to intact underlying knowledge of language—have largely determined the major approaches to treatment. The "loss" theory has led to pedagogic approaches, while the "impaired access" theory has led to stimulation and reorganization approaches. Since treatments of various dysphasic syndromes can be viewed in the light of either theory, these views will determine the protocol used here for examining some of the more robust approaches to treatment of some major dysphasic syndromes.

## 2. Treatment of Global Dysphasia

Globally dysphasic patients are severely impaired in all language modalities, so that they have little ability to communicate through speech or writing, and little ability to comprehend either the spoken or the written word. In this respect they present the greatest rehabilitative challenge. After concentrating an impressive amount of time, effort and innovative skill towards treating such patients, Schuell, Jenkins, Jiménez-Pabón (1964) concluded that any gains made by these patients while in treatment do not become functional. They are never able to use language to help them carry out plans or intentions, and therefore, have an "irreversible syndrome". In fact, these authors felt that it was not sound practice to treat such patients for months or years expecting that they will regain functional language skills.

Despite this gloomy picture clinicians persist in their efforts to treat global dysphasics, in the hope that some approach will prove effective.

### Specific Approaches to Global Dysphasia

*Teaching Language to Global Dysphasics*

As stated earlier dysphasia may be viewed as 1) the loss of language or 2) the impaired ability to gain access to language. According to the former view, global dysphasia represents a "massive language loss" (Glass, Gazzaniga, and Premack, 1973). Since global patients are unable to use successfully any form of natural language one might assume that the extensive left hemisphere lesion usually associated with this syndrome has effectively erased nearly all linguistic knowledge. If this is the case, then two options present themselves: 1) One could try to rebuild or reteach linguistic skills from the ground up or 2) One could set natural language aside, and attempt to teach an alternate form of communication. The former approach is historically most prevalent. Froschels (1933) called this method "brain gymnastics" because it involves massive amounts of drill, with the patient repeating and writing individual sounds, then learning to repeat two connected syllables and then real words and so on. Since there is no good evidence that language is acquired in this manner developmentally (Menyuk, 1971), this approach would not constitute a rebuilding program in the developmental sense. In addition, there is good evidence that dysphasics do better with meaningful units than with nonsense syllables and with more, rather than less information. Geschwind (1965), for example, has observed that severe dysphasics retained a disproportionate capacity for responding to whole body commands such as "stand up". Johnson et al. (1976) compared the ability of patients whose standardized test scores classified them as global patients, to follow whole body commands and to discriminate auditorily single nouns such as "pen". The results of this study confirmed Geschwind's observation. Johnson et al. concluded their paper with the suggestion that whole body commands might be a basis for treatment of global patients—a suggestion made nearly 40 years earlier by Backus (1937).

There have been several attempts to teach hearing dysphasics to use the manual sign languages used by the deaf. This approach may be particularly appealing with global patients, who seem to have no appreciation or use of the spoken word. Although there are reports (Eagelson et al., 1970; Chen, 1971) of the successful use of signing with other types of dysphasics, we could find no published data indicating that global patients can acquire functional signing skills. This is not surprising, since among other problems these patients usually have hemiplegia, and may be forced to use the non-dominant (often apraxic) hand. Dyspraxia refers here to "the incapacity for purposive movement of the limbs despite retained mobility" (Liepmann, quoted in Geschwind, 1967).

Instead of instructing global patients in the use of manual sign language, at least two teams of investigators have chosen to train artificial language in which shapes or line drawings stand for words or concepts. The first such study was undertaken by Glass, Gazzaniga, and Premack (1973), who set about training globally dysphasic patients to use an artificial language system using cut out paper symbols. These symbols could be arranged to express various relationships which included same/different, negation, and subject-predicate-direct object statements of action. They found that patients attained varying levels of competency in expressing these relationships with this symbol system. These authors felt that because globally dysphasic patients still have the ability to learn an artificial language, they retain a rich conceptual system despite a massive language loss, and that global patients do not suffer cognitive impairment in direct proportion to their language impairment.

This opinion is supported by the results of a recent study (Risse, 1978) in which 20 dysphasic patients were asked to perform various Piagetian type developmental problem solving tasks. No relationship was found between severity of dysphasia and conceptual performance. In fact, two global dysphasics demonstrated almost totally intact conceptual performance.

An interesting finding of the Glass, Gazzaniga, Premack study is that with few exceptions their globally dysphasic patients were able to complete the pretest which required them to form a sorting hypothesis and class pictures as animate/inanimate, indoors/outdoors and so forth. This is reminiscent of an observation made by Wepman in 1951. He found that when one of his global patients was left alone with geometric shapes he spontaneously sorted them according to color. This proved to be the starting point for a new course of treatment. Wepman began training color naming within the confines of the patient's articulatory skills. After several intermediary steps the patient was eventually treated as an "expressive" dysphasic.

Thus, there is some good evidence that globally dysphasic patients do possess certain basic logical capacities. This will come as no surprise to clinicians who have observed "global" patients who are perfectly well oriented to their surroundings and are able to appreciate most of the nonverbal nuances of their environment. The primary goal of language rehabilitation, however, is to improve functional communication, so dysphasiologists continue to explore approaches which hold communicative promise for global dysphasics. Baker et al. (1975) and Gardner et al. (1976) used a visual communication system which utilized a series of index cards, each containing a simple arbitrary (geometric) or representational (ideographic) form, denoting a meaningful unit, to investigate 1) whether the cognitive operations entailed in natural language persisted after natural language capacities are destroyed and 2) whether severely dysphasic patients could be taught to communicate using an alternative or visual symbol system. Eight globally dysphasic patients were studied and attained varying levels of visual communication, which involved 1) carrying out commands 2) answering questions 3) describing events 4) expressing feelings 5) expressing immediate needs 6) expressing desires. These findings indicate that some global patients can master the basics of an alternative symbol system, and that at least some of the cognitive operations necessary for natural language are intact despite the severe dysphasia.

*Re-Establishing Access to Language in Global Dysphasics*

The above observation suggests a second view of global dysphasia, not as a "massive loss of language" but as a severe impairment of the ability to gain access to what is known about language. In addition to the formal evidence that the latter view may be more appropriate and descriptive of the problem, there is the informal clinical evidence that with the proper cues, prompts and stimulation, even the most impaired dysphasic patients can both comprehend and produce speech. But while these clinical aids temporarily help the patient to tap his linguistic store, they apparently do not improve self—initiated retrieval skills. In fact, there is currently no published technique which succeeds in accomplishing this task, although a recent pilot treatment study undertaken with globally dysphasic patients at the Boston Veterans Medical Center has proven hopeful. The treatment is called Visual-Action Therapy (V.A.T.) (Helm and Benson, 1978). This method trains the patient to associate ideographic forms with particular objects and actions and to carry out a series of tasks in association with these drawings. V.A.T. employs 8 real objects, such as cup and razor, all of which can be easily manipulated with one hand and symbolized by a distinct gesture. The program is hierarchally arranged into steps and levels which are aimed at gradually shaping the patient first to appreciate the symbolic value of representational line drawings and gestures and later to produce symbolic gestures. To date approximately one dozen globally dysphasic patients have received V.A.T. training. Comparisons of pre- and post-treatment standardized test scores showed all patients improved most notably in the areas of auditory comprehension and gestural pantomime. Improvement in other selected language skills such as naming and writing was also seen among individuals. These results were obtained despite the fact that the entire V.A.T. program is conducted in silence. This finding supports the view that language may not be destroyed in global dysphasia, but instead may no longer be accessible via primary pathways.

## 3. Treatment of Non-Fluent Dysphasia

While most dysphasiologists would agree that severe impairment in all language modalities following brain damage should be called "global dysphasia", few dysphasiologists agree on the term for the acquired disorder in which output is sparse, effortful and dysprosodic, while auditory comprehension is relatively spared. Johns and LaPointe (1976) make this point effectively.

"It can be seen that such terms as aphemia, Broca's aphasia motor dysphasia, predominantly expressive dysphasia, subcortical motor dysphasia, anarthria, verbal dysphasia, phonetic disintegration of speech, apraxia, apraxic dysarthria, cortical dysarthria, and oral verbal apraxia, while attempting to describe a particular set of speech behaviors, can be extremely confusing and bewildering to the clinician who is faced with designing appropriate therapeutic procedures for such a patient." (p. 163)

Since all of the above mentioned disorders are forms of non-fluent dysphasia, the use of this general rubric allows us to discuss a variety of treatment approaches which have been associated with specific forms of non-fluent dysphasia.

*Non-fluent dysphasia* is a disorder in which verbal output is sparse, and effortfully articulated. Production often proceeds in a syllable by syllable fashion, so that the melodic line of speech is lost or impaired. Phrase length is reduced, so that utterances often consist of single words. Substantive words predominate so that utterances have an agrammatic or telegraphic quality. Verbal or semantic paraphasias in which one real word is substituted for another are rare. Auditory comprehension is often remarkably well preserved, although moderate impairment in carrying out sequential commands may exist (Benson, 1967; Goodglass and Kaplan, 1972). Patients having non-fluent dysphasia often have bucco/facial dyspraxia (De Renzi, Pieczuro and Vignolo, 1966), and right hemiparesis (Benson and Geschwind, 1971). Brust *et al.* (1976) found that of 177 acutely dysphasic patients admitted to Harlem Regional Hospital, 68 % had non-fluent dysphasia.

In its severest form non-fluent dysphasia is somewhat paradoxical in that although patients have an inability to articulate much beyond a restricted phonemic, syllabic, stereotypy such as, "bika-bika", these syllables are produced over and over with apparently good articulation and melody. If the patient recovers some real speech, however, the amelodic quality and effortful articulation typical of non-fluent dysphasia becomes manifest. These patients present an enormous rehabilitative challenge. In fact, Vignolo (1964) states that presence of a restricted phonemic syllabic stereotypy may be a bad prognostic sign.

Severe non-fluent dysphasia can be viewed either as the loss of knowledge of expressive speech, or the inability to gain access to the still intact knowledge of expressive speech.

## Specific Approaches to Non-Fluent Dysphasia

*Teaching Expressive Speech Skills*

The loss of knowledge view of non-fluent dysphasia was apparently held by Froschels (1933), who stated that such patients are like deaf and dumb individuals who are deficient in the collective sounds of speech. In Froschels' opinion the patient must be taught to speak again, sound by sound, syllable by syllable, until real words and eventually sentences can be introduced. This hierarchy also includes certain phonological considerations. For example, one begins with the most visible sounds such as bilabial plosives (b, p) and the open vowel (ah). This treatment, which sometimes includes the actual physical manipulation of the articulators with a spatula and stylet, often takes place in front of a mirror. Froschels also recommended that the clinician incorporate into the practice procedure any sound that the patient accidently produces.

*Practice* is a crucial factor in the motor placement approach to severe non-fluent dysphasia. This point was quite forcibly expressed by Granich (1947), who stated that the clinician must carry on persistent taxing drills with the patient. For more complete descriptions of the sound building, motor placement approach the reader may also look to Corbin (1951); Schuell, Carroll, and Street (1955); Dabul and Bollier (1976).

One severely non-fluent patient for whom this sound building, motor placement approach seems to have had positive effect is described by Singer and Low (1933). They reported the case of a woman who for two years had persistent severe hemiplegia and dysphasia following a stroke. Although "she understood everything that was said to her" her only verbal response was "o-de-dar". These syllables were uttered rapidly and repeated frequently.

Singer began by helping her combine the sound "ee" with a blowing sound to produce "bee". Then she was taught "ba" then "bar". Subsequently, the patient's husband helped her in daily practice. The consonant sounds were trained first. Following this, each consonant was combined with various vowels. Then terminal consonants were added to produce such words as "bed" "bowl" and "book". Finally, syllables were combined to produce such words as "table" "window". Within a few months the patient was able to articulate a large number of words. The patient was seen eight years later and found to have a vocabulary of 500 words, some of which were difficult for strangers to understand.

### Re-Establishing Access to Expressive Speech Skills

Only the strongest skeptic would question that the sound building, motor placement approach used with Singer and Low's patient described above was effective. The patient had produced nothing but a restricted phonemic stereo-type for two years and then within months after beginning daily treatment she was producing real words. One is struck, however, with the limitation of verbal expression (500 words) which still existed eight years later, partic-ularly since "the training was continued with the same assiduity" during this period. The fact that this patient progressed from a nonsense stereotypy to real word production within a few months time lends support to a loss of access view of dysphasia, as the response to treatment seems too rapid to be attributed to learning. One might also assume that if her speech gains were the result of learning, she would continue to learn more from the lessons that continued over the next eight years. Instead she apparently plateaued early in the treatment programm, suggesting that an inhibiting force was once again present. Interestingly enough, many such patients are able to articulate a wide range of phonemes within certain words and not within others. This suggests that the knowledge of sounds is not lost, but merely that they are inconsistently accessible. It is as if these patients are sometimes prevented or in-hibited from speaking what they know.

Clinicians who hold this view have developed various facilitory techniques geared to disinhibit verbal expression in non-fluent dysphasics. The patients are not taught sound production. Instead, they are stimulated toward pro-duction of real speech via some facilitory technique.

Two such methods currently in use are: Melodic Intonation Therapy (Sparks, Helm, Albert, 1974) and Amerind (Skelly et al., 1974). The con-cepts which form the basis of both of these techniques may be found in Backus (1945), who was mentioned earlier as a proponent of the "intact knowledge" theory of dysphasia (Backus, 1937). She stated that speech cannot be built up sound by sound. Instead, speech patterns must be stressed often through singing, and rhythmic unison speech. These statements are appropriate to the Melodic Intonation Therapy approach which was formally developed and investigated more than thirty years later.

The impetus for the Melodic Intonation Therapy study was the clinical observation that some severely dysphasic patients can sing better than they can speak. In fact Mills (1904) suggested that clinicians would do well to sit down at the piano with such patients and sing popular songs. Many institutions have heeded this advice and use group singing as part of their rehabilitation program. Although singing popular songs may benefit the patient psychologically, it apparently has little effect on production of propositional speech.

For this reason Albert, Sparks, and Helm (1973) decided to explore the possibility that some alternative form of singing might be used to improve the speech of long term severely dysphasic patients. The increasing evidence that the right cerebral hemisphere is dominant for music (Spellacy, 1970) and intonational contours (Blumstein and Cooper, 1974) lent further support to this proposed approach. These findings suggested that the functions which are associated with the presumably intact right hemispheres of these patients might be used to improve the language functions of their damaged dominant left hemispheres. The program which evolved utilizes a large number of high probability phrases and sentences such as, "Open the door", which are sung instead of spoken. These linguistic items are embedded in a multilevel hierarchy, which requires the clinician to withdraw support gradually while the patient is required to demonstrate an increasingly independent level of verbal initiation. For example, the first level of the program includes the following five steps.

1. The clinician begins by singing (intoning) the target phrase to the patient. No response is required.

2. The clinician and the patent sing the phrase in unison.

3. The clinician and the patient sing the initial part of the phrase in unison, but at about the halfway point the clinician fades out and the patient is required to complete the phrase alone.

4. The patient repeats the sung phrase after the clinician.

5. The patient sings the target phrase in response to a probe question which is sung by the clinician.

Throughout this routine the clinician holds the patient's hand and taps out each syllable in a rhythmic fashion. Each response is scored so that the patient's progress within the program is assessed by objective criteria (Sparks, Helm, Albert, 1974; Sparks and Holland, 1976). The "practice effect" which is a vital part of pedagogic approaches is avoided in this treatment through the use of a changing variety of phrase items within and between sessions.

In planning the Melodic Intonation Therapy program the investigators used a programmed instruction approach which applies principles of shaping and differential reinforcement to learning tasks. One begins writing such a program by defining the ultimate behavior desired. In the case of Melodic Intonation Therapy the ultimate behavior desired for severely non-fluent dysphasic patients is to help them become functional speakers. To this end the final step of the multilevel program requires the patient to produce normally spoken target phrases in response to normally spoken probe questions. The method has proven successful with some severely non-fluent patients in so far as their naming capacity and conversational phrase length has increased with treatment. Such a patient is described below.

# A Case Report of Treatment with Melodic Intonation Therapy

Mr. H., a forty-one year old right-handed bricklayer with a 10 year history of diabetes, hypertension and obesity, was first seen at the Boston Veterans Administration Hospital in June, 1972. One month previous to that admission he developed sudden onset of right-sided weakness and difficulty in speaking without loss of consciousness. He was taken to a local general hospital where neurological examination showed dense flaccid right hemiparesis, questionable homonymous hemianopia, left conjugate deviation of the eyes and bilateral Babinski reflexes.

Mental status examination showed Mr. H. to be awake, alert and co-operative, with a tendency to cry when frustrated by his inability to communicate.

Informal dysphasia examination showed that spontaneous speech was limited to a few stereotyped phonemes ("Dis, dis, kon, dis"). Auditory comprehension was impaired for "yes"/"no" questions and pointing to objects. There was a tendency to perseverate. He was, however, able to perform one-step whole body commands. Repetition, like spontaneous speech, was limited to the phonemic stereotypes, as was serial speech and naming. Singing was poor. Writing was limited to a few letters written to confrontation. He was unable to read aloud, but was able to point out a few printed letters, words and numbers.

Related cortical function testing showed severe bucco-facial apraxia, even upon imitation. Limb praxis was poor to command but improved upon imitation.

The brain scan showed a perisylvian lesion extending frontally and a separate parietal lesion extending deeply.

In addition to the informal language evaluation Mr. H. was tested with the Boston Diagnostic Aphasia Examination (Goodglass and Kaplan, 1972). The results indicated global impairment in all modalities except auditory comprehension.

During the following three months Mr. H. was seen for daily language rehabilitation. This took the form of auditory training and other traditional dysphasia treatment tasks, such as cued speech. Little improvement was noted in either speech output or auditory comprehension, reading or writing skills, beyond the ability to repeat a greater variety of simple phonemes. His hemiparesis, however, had improved so that the patient was now able to walk with only mild difficulty and had regained limited use of the right arm and hand. The patient's diabetes was under control and he had lost 34 lbs. He was discharged home where he lived with his wife and four children.

During the next five months, Melodic Intonation Therapy was developed and used to treat three patients at the Boston Veterans Administration Hospital dysphasia unit. Mr. H. was considered to be a possible candidate because he met the criteria we had set up for admittance to that study, that is 1) he had not improved in language function for at least four months despite language rehabilitation; 2) he had severe paucity of speech output; 3) his auditory comprehension was no longer severely impaired. He was re-admitted in late March,

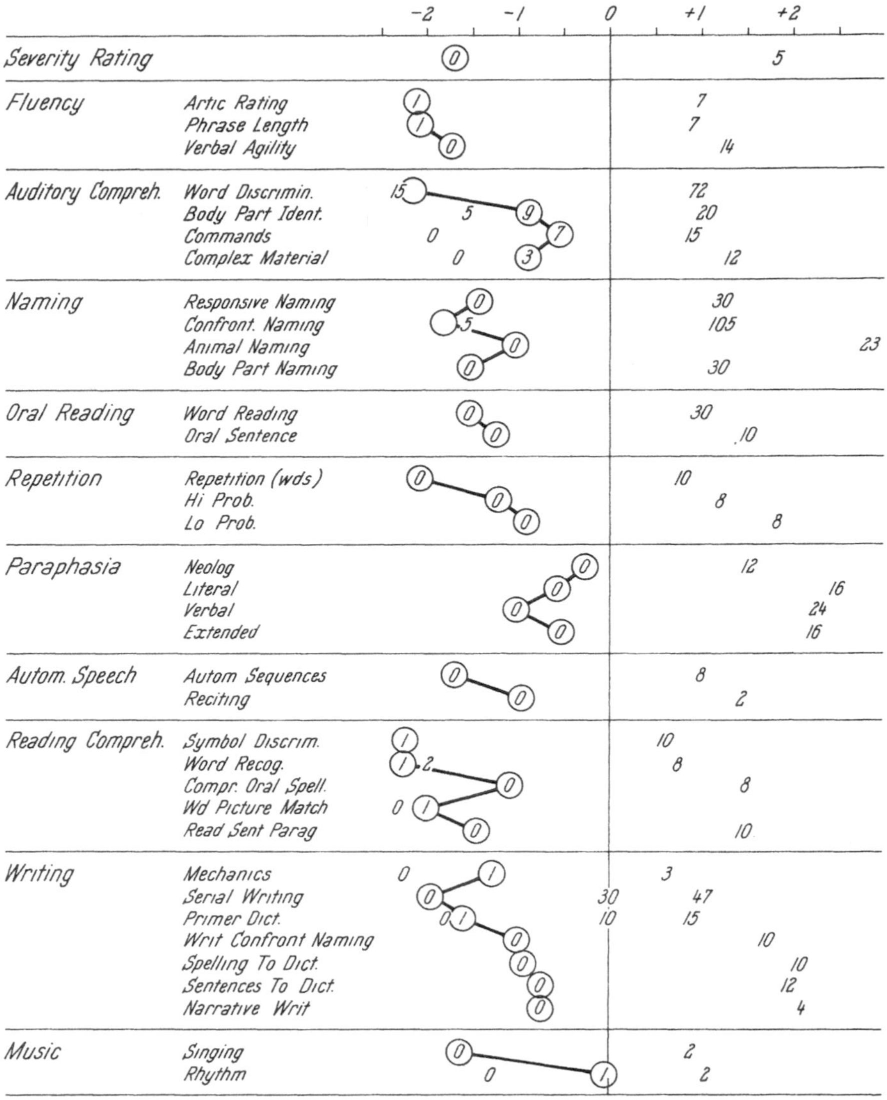

| | | −2 | −1 | 0 | +1 | +2 |
|---|---|---|---|---|---|---|
| Severity Rating | | 0 | | | 5 | |
| Fluency | Artic Rating | 1 | | | 7 | |
| | Phrase Length | 1 | | | 7 | |
| | Verbal Agility | 0 | | | 14 | |
| Auditory Compreh. | Word Discrimin. | 15 | | | 72 | |
| | Body Part Ident. | 5 / 9 | | | 20 | |
| | Commands | 0 / 7 | | | 15 | |
| | Complex Material | 0 / 3 | | | 12 | |
| Naming | Responsive Naming | 0 | | | 30 | |
| | Confront. Naming | 5 | | | 105 | |
| | Animal Naming | 0 | | | | 23 |
| | Body Part Naming | 0 | | | 30 | |
| Oral Reading | Word Reading | 0 | | | 30 | |
| | Oral Sentence | 0 | | | 10 | |
| Repetition | Repetition (wds) | 0 | | | 10 | |
| | Hi Prob. | 0 | | | 8 | |
| | Lo Prob. | 0 | | | | 8 |
| Paraphasia | Neolog | 0 | | | 12 | |
| | Literal | 0 | | | | 16 |
| | Verbal | 0 | | | | 24 |
| | Extended | 0 | | | | 16 |
| Autom. Speech | Autom Sequences | 0 | | | 8 | |
| | Reciting | 0 | | | 2 | |
| Reading Compreh. | Symbol Discrim. | 1 | | | 10 | |
| | Word Recog. | 1 2 | 0 | | 8 | |
| | Compr. Oral Spell. | | | | 8 | |
| | Wd Picture Match | 0 1 | 0 | | | |
| | Read Sent Parag | 0 | | | 10 | |
| Writing | Mechanics | 0 1 | | 3 | | |
| | Serial Writing | 0 | | 30 | 47 | |
| | Primer Dict. | 0 1 | | 10 | 15 | |
| | Writ Confront Naming | 0 | | | 10 | |
| | Spelling To Dict. | 0 | | | | 10 |
| | Sentences To Dict. | 0 | | | | 12 |
| | Narrative Writ | 0 | | | | 4 |
| Music | Singing | 0 | | | 2 | |
| | Rhythm | 0 / 1 | | | 2 | |

Fig. 10. Z-score profile of dysphasia subscores. Name: T. H. Date of examination: June 15, 1972

1973 for re-evaluation of his dysphasia and a course of Melodic Intonation Therapy.

Medical examination demonstrated a right hemiparesis and hemi-sensory loss. The patient was re-evaluated with the Boston Diagnostic Aphasia Examination. The scores, which are reported below, show that a considerable improvement in the auditory discrimination section of auditory comprehension had occurred during the six months at home. There was no change in speech scores and spontaneous output was essentially unchanged and remained narrowly stereotypic "dis a kom adis a ka".

He was started on a daily regime of Melodic Intonation Therapy, amounting to approximately 10 sessions per week. Within two weeks Mr. H. was observed by the nursing staff to use some meaningful single syllable words.

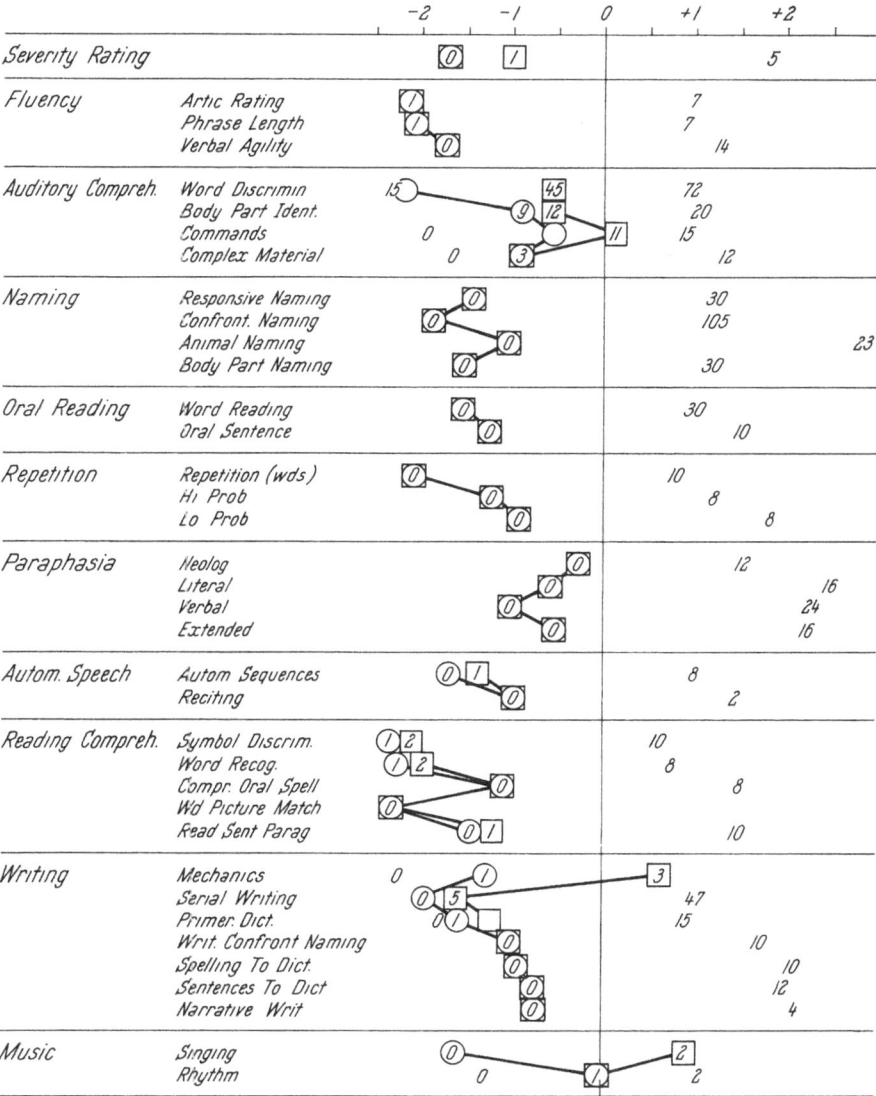

Fig. 11. Z-score profile of dysphasia subscores. Name: T. H. Date of examination: ○ June 15, 1972; □ March 23, 1973

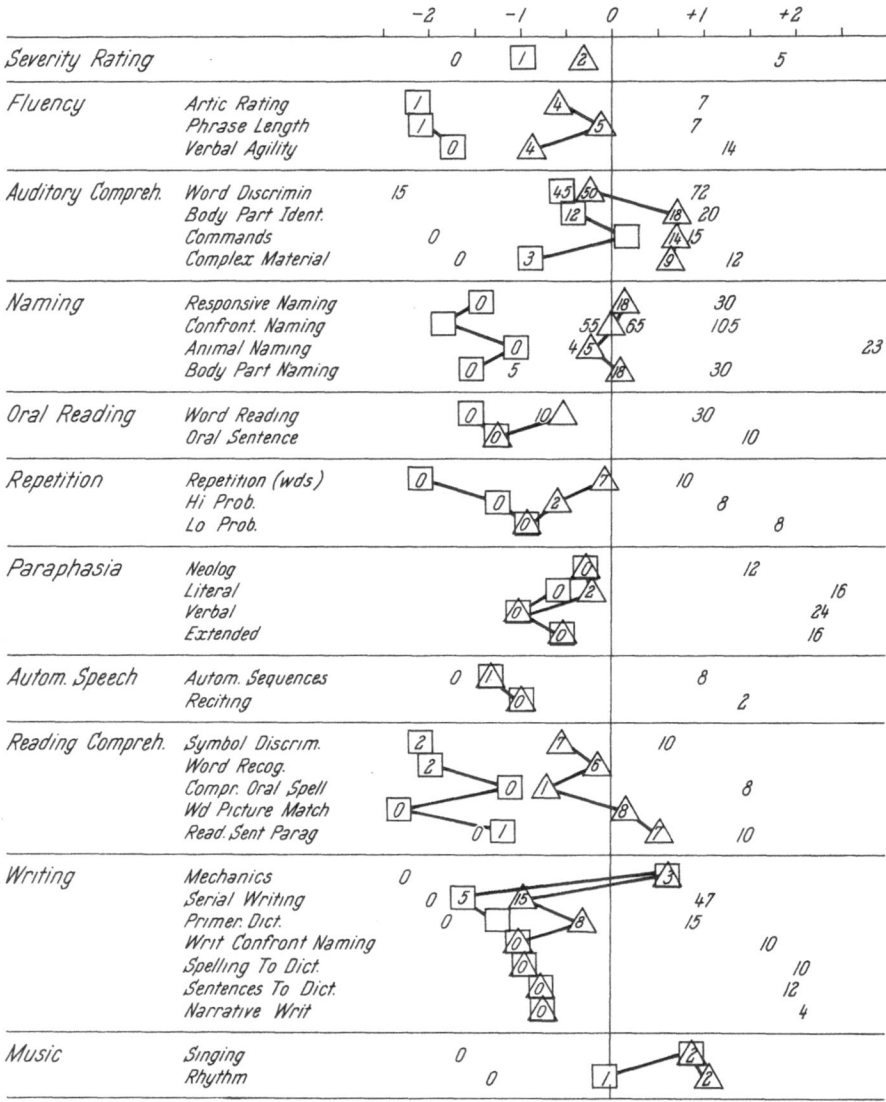

Fig. 12. Z-score profile of dysphasia subscores. Name: T. H. Date of examination: □ March 23, 1973; △ July 17, 1973

Following approximately 120 sessions of M.I.T., Mr. H. was re-evaluated. These post-melodic intonation therapy scores are contrasted with pre-melodic intonation therapy scores (Fig. 12). They show that with this new treatment Mr. H. experienced a breakthrough in verbal skills and reading comprehension for the first time since his stroke some nine months before. Since melodic intonation therapy is designed to improve expressive speech skills, the method was considered successful. But of equal interest is the fact that reading skills improved dramatically, even though no written stimuli are used in this method. This finding suggests that dysphasia treatment may improve retrieval skills rather than "teach" the patient language.

Although test scores are important to measuring progress, it is perhaps more important to note that Mr. H. was now able to communicate, albeit somewhat dysarthrically, with those in his environment, making his wants and opinions known. Brief transcripts of pre- and post-melodic intonation therapy conversation llustrate this point.

## Pre-Melodic Intonation Therapy Speech

Examiner: "How are you?"
Patient: "goo-go."
Examiner: "What is your full name?"
Patient: "Doe-kah."
Examiner: "What have you been doing?"
Patient: "Dis ah dis."

## Pre-Melodic Intonation Therapy Speech

Examiner: "How are you?"
Patient: "All right."
Examiner: "Where have you been?"
Patient: "Dis week-end-we go home."
Examiner: "What did you do?"
Patient: "I work."
Examiner: "What kind of work?"
Patient: "Car-pen-try."
Examiner: "Did you go out Saturday night?"
Patient: "Po-dish cub."
Examiner: "Polish Club? What went on there?"
Patient: "A dance."

The gains made during the new treatment were not transitory, as demonstrated in the following brief transcription taken from a conversation 6 weeks after discharge.

## Six Weeks Post-Melodic Intonation Therapy Speech

Examiner: "What brings you here?"
Patient: "I come to see a friend."
Examiner: "How did you get here?"
Patient: "I came-here-by-car."
Examiner: "Did you drive?"
Patient: "No, an-other friend."

In addition to suggestions as to the use of rhythm and melody for improving speech output, Backus (1945) suggested that clinicians use pantomime activities with dysphasics, because speech is a total body response. These activities can incorporate unison speaking, imitation, delayed imitation and finally evocation of the oral response from memory at the appropriate time.

These statements very simply describe the Amerind (American Indian Sign) approach which was formally researched decades later, and found to be an effective approach to treatment of severely non-fluent dysphasic patients. Skelly *et al.* (1974) described six patients who were treated according to this method, which combines gestural sign and oral speech production. Patients were encouraged to produce both specific gestures and their verbal counterpart first in unison with a letter with only slight cues from the clinician. During six month training periods which were initiated at least two years after onset of dysphasia, three of the six patients progressed from no usable speech to three word sentences, one patient progressed from no usable speech to two word sentences, and one from no usable speech to single word usage. The sixth patient remained unable to speak intelligibly.

Amerind lends itself well to group treatment, so that an Amerind group of both in- and outpatients presently meets at this institution. Furthermore, some inpatients are able to "practice" sign production individually via a Polavision player. This newly developed instrument has been made available to us for experimental work. Short films which demonstrate approximately 8 gestural signs which are related to pictures (*e.g.*, a picture of a house) are played by the patients themselves in our treatment laboratory. As the clinician on the film makes a "sign" the patients initiate that sign. More advanced patients can produce the proper sign after merely seeing the pictured stimuli which come first and use the clinician's signing as a re-inforcement.

Both Melodic Intonation Therapy and Amerind may be viewed as facilitory approaches which allow the patient to gain access to what he knows about expressive speech via alternative pathways.

# 4. Treatment of Agrammatism

The case study which was presented above serves to illustrate two points. The first is that patients who have remained severely impaired in verbal expression past the period of natural spontaneous recovery can regain functional speech skills with special treatment. The second is that while the verbal expression of such patients may improve significantly under treatment, the improved output may be characterized as agrammatic, in so far as conversational phrase length and syntactical forms are restricted. Goodglass (1976) distinguished two types of dysphasic grammatical disorders, motor agrammatism and paragrammatism. Since we are concerned here with the variety of agrammatism associated with non-fluent dysphasia, the following deals with the disorder known as motor agrammatism in which articles, connective words, auxiliaries, and inflections tend to drop out, leaving content words such as nouns, verbs and adjectives.

In keeping with the "therapeutic models concept" which has been adopted for this chapter, agrammatism may be viewed as a loss of syntactic knowledge or as reduced efficiency in gaining access to that knowledge. If agrammatism is viewed as a loss of the knowledge of syntax then the task is to replace the lost knowledge by teaching syntactical rules according to either theories of normal language acquisition or approaches used in teaching English as a second language to adults. If agrammatism is viewed as a reduction of efficiency in gaining access to syntactical knowledge then the task is to devise techniques which will somehow facilitate new accessing strategies. In accomplishing this task the clinician may rely upon neurolinguistic studies which examine the nature of dysphasic language disturbances.

The literature on dysphasia provides support for both the "loss" and the "reduced efficiency" models of dysphasic agrammatism. Some dysphasiologists feel that agrammatism reflects linguistic limitations (Zurif et al., 1976). For these authors agrammatism represents at least a partial loss of tacit knowledge as evidenced by performance on reading tasks. Other dysphasiologists propose that the tacit knowledge of language is spared in agrammatic patients as evidenced by their preserved auditory comprehension (Weigl and Bierwisch, 1970; Lenneberg, 1973). These authors view agrammatism as an economizing measure to circumvent the articulatory impairment.

Let us now consider treatment according to the view that agrammatism represents a loss of syntactic knowledge.

*Reteaching Syntax*

Three clinicians who treat agrammatism as if it were a loss of "syntactic structures" are Crystal, Fletcher, and Garmin (1976). While these authors make no claims as to how these structures come to be lost, they believe that developmental norms should be used to determine the order in which structures are introduced in rehabilitation. Since a full chapter of their text is dedicated to a detailed description of this approach as applied to one patient, it will not be reproduced here; instead the reader is encouraged to examine this chapter, as it is a good example of the reteaching approach, which begins with elicitation of clause-level elements such as subject, verb + object, and later introduces such elements as adjectives, adverbs, prepositions, negatives, and conjunctions.

Two studies which take a non-developmentally based learning approach to rehabilitating agrammatism are described by Holland and Levy (1971) and Naeser (1975). Holland and Levy (1971) used a programmed instruction procedure to train dysphasic subjects to use an active sentence, for example, "The woman opens the door". This sentence type was chosen because of studies which suggest that it is the easiest for both normals and dysphasics to understand. These investigators found that although training resulted in improved ability to use active sentences, there was limited generalization to use of interrogatives, and no generalization to negatives, passives, and 4 sentences which incorporated lexical changes in the trained active sentences.

Naeser (1975) also used a programmed instruction approach to train dysphasic patients to produce three basic declarative sentence types. Type one used the verb form "to be", for example "that is a house". Type two used a transitive verb, for example "The woman opens the door", and type three used an intransitive verb, for example "soldiers march". Subjects showed an average improvement of 10 % in their ability to produce the trained basic sentence types. There was 36 % improvement in pattern carryover to untrained sentences of the same type. In addition, the subjects improved 30 % in their ability to produce subject/verb agrements, in which the verb number agrees with the subject.

Again, the real test of whether any rehabilitation approach works is whether there is improvement in functional speech. In the case of the agrammatic patient we must ask whether our techniques are effective in making the patient less agrammatic as measured by the variety of syntactical forms used during day by day verbal exchanges. This yardstick should be used to measure the above techniques and those described below which strive to treat agrammatism according to the view that it represents a retrieval disorder.

### Re-Establishing Access to Syntax

Unlike those who view agrammatism as a loss of syntactical knowledge, Weigl and Bierwisch (1970) feel that agrammatic dysphasic patients retain their tacit knowledge of syntax. These authors speak of dysphasic syndromes as representing disorders of competence while language remains intact. This view is based in part on the variability of performance which is seen among dysphasics. Furthermore, these authors feel that the blocks which interfere with accessing strategies can be removed with the proper cues. This "deblocking" process, when applied to syntax, takes the form of introducing the patient to all the possible tense changes of a particular verb embedded in predetermined carrier sentences, for example "the mother *irons* the shirt", "the mother *ironed* the shirt", "the mother *will iron* the shirt", etc. Further description of the "deblocking" phenomenon may be found in a 1968 essay by E. Weigl.

A second approach to stimulating access to syntactical knowledge has been proposed by Helm (1977). This approach is based upon the results of two studies undertaken by colleagues at the Boston Veterans Administration Medical Center (Goodglass *et al.,* 1972; Gleason *et al.,* 1975). In these studies a story completion technique was used to elicit verbally 14 different English syntactic constructions from agrammatic patients. An example of this technique is as follows: (Examiner): "My friend comes in. I want him to sit down. So I say to him, what?" (Patient): "Sit down". Although an order of difficulty exists, agrammatic patients were able to produce a variety of constructions within highly structured linguistic environments. The investigators concluded that syntactical knowledge is not lost, instead, the dysphasic has an impaired and inconsistent ability to gain access to that knowledge.

The story completion technique described above is used in the syntax stimulation program currently undergoing investigation in our unit. This program elicits multiple examples of the five sentence types which were found to be the easiest for agrammatic patients to produce in a story completion task (Gleason *et al.,* 1975). The order of presentation follows the order of difficulty discovered by these investigators, since it is axiomatic in dysphasia rehabilitation that one begins where the patient has the greatest chance of success, and progresses in small steps. Furthermore, since there is evidence that a phonological hierarchy of difficulty exists for non-fluent dysphasic patients (Alajouanine, Ombredane, and Durand, 1939; Hatfield, 1972) and that phonological structure influences sentence production (Martin, 1977), this program controls for phonology, so that the easier sentences are made up of words which are also phonologically easier. For example, "sit down" is a choice for sentence type 1 (imperative intransitive), but not "stand up", because consonant blends are known to be particularly difficult for non-fluent dysphasic patients.

The syntax stimulation program uses a story completion technique to elicit twenty samples each of the following five sentence types:

| Sentence Type | Examplar |
|---|---|
| 1. Imperative intransitive | "Wake up." |
| 2. Imperative transitive | "Wash the dishes." |
| 3. WH Interrogative | "What are you reading?" |
| 4. Declarative transitive | "He rings the bell." |
| 5. Declarative intransitive | "He swims." |

Although this method is still under formal investigation, the preliminary findings are encouraging in so far as patients may show improved ability to produce grammatical strings in conversational and expository speech as measured by the Boston Diagnostic Aphasia Examination. It is felt that the use of the story completion technique accompanied by appropriate visual stimuli increases the chances of a patient generating a correct response, because visual as well as verbal imagery is heightened (West, 1977).

## The Preventive Approach to Agrammatism

A unique and reportedly effective approach to treatment of agrammatism has been described by Beyn and Shokhor-Trotskaya (1966). Instead of trying to correct the telegraphic style after it has been manifested, these clinicians tried to prevent this style from developing by stimulating severely impaired patients to produce non-nominative holophrastic utterances at an early stage in the recovery process. Twenty-five non-fluent patients, one to three months post ictus, were encouraged through unison speech and repetition tasks to use such words as "oh", "no", "there", "now", "give", "good". Later such expressives as "I want" were combined with verbs such as "to eat", "to sleep". The names of objects were withheld until they appeared in spontaneous speech and even then were not introduced in the Russian nominative case. The authors report that none of the 25 patients developed telegraphic speech, compared with patients who were traditionally treated. Unfortunately, the group treated preventively apparently had no "apraxia of articulation" and could not be considered "cortical efferent motor dysphasics" (Luria's term), so that we must question whether they would ever have been diagnosed as true non-fluent agrammatics. Certainly, however, the use of holophrastic speech in the treatment of severe dysphasia is worth considering. As Beyn and Shokhor-Trotskaya point out, the speech pathologist may actually contribute to the patient's agrammatism by training object names and using a telegraphic style with these patients.

# 5. Treatment of Anomia

It is understood that virtually all dysphasics have some disturbance of naming, which can be demonstrated by depressed raw scores on confrontation naming tests. The defect underlying the naming disorder, however, differs according to type of dysphasia or site and extent of lesion.

Goldstein (1942) described two varieties of naming disorders. One involves loss of the "abstract attitude", that is, the patient no longer associates words as symbols of the things they name. This variety of naming disorder contrasts with a second naming disturbance resulting from a loss of the "instrumentalities of speech" which impairs the patient's ability to utter the word. Head (1926) described a disorder he called "nominal aphasia" which represents an inability to recognize words as symbols for objects, actions, colors, etc.

Luria (1966) described two varieties of naming disturbances. One involves a defect of selectivity at a high level of semantic organization and word choice. The other involves a defect in modality specific sensorimotor processes of output. Geschwind (1967) also described four varieties of naming errors. In one, dysphasic anomia, the patient has disturbance in confrontation naming. In the second, the patient is able to perceive stimuli but because of major pathway disruption, the information is not transmitted to the language area for naming. The third is called non-dysphasic misnaming and is seen in patients without obvious dysphasia. The fourth, hysterical misnaming is seen in patients mimicking dysphasia.

Goodglass et al. (1976) contrast the verbal amnesic who misnames because he has lost the association between word and concept with the Broca's dysphasic who has tacit knowledge of the word but misnames because he has lost the ability for its motor realization.

This section is devoted to treatment of that variety of anomia which involves impairment of the meaning of words, the defect of semantic selectivity, the inability to associate word and concept. Although anomia may represent a relatively milder dysphasic condition, it does not respond easily to treatment. Luria (1970) states that all disturbances of nominative function are extremely difficult to overcome. Kertesz and McCabe (1977) studied the recovery patterns and prognosis of 93 dysphasics. They found that although anomic dysphasics tended to have the mildest disturbances, some such patients continued to have long term naming problems which may prevent employment. Thus, the anomic dysphasic presents a distinct challenge to the language clinician. The treatment approach which the clinician chooses or devises, however, will be determined by his view as to the nature of the disorder known as anomia.

*Reteaching the Names of Things*

Anomia may be viewed as a reduction of the lexical store, that is, the loss of the names of things. The clinician who holds this view will undertake tasks which are aimed at helping the patient "relearn" names, often through the use of pictures of objects (Luchsinger and Arnold, 1965).

Wepman (1951) took a "Pavlovian approach" to establishing names as substitutes for objects. He recommended that the clinician concentrate on a few words, which are presented over and over again for the patient to see, hear, repeat, trace, and read. A case was cited in which during 3 months of teaching four words, the patient learned two words, but then within two weeks he made rapid progress, so that by 5 months he was "almost better". Unfortunately, the case history was too scanty to allow us to weigh the effects of treatment, against those of spontaneous recovery. This case does, however, make an interesting point, one also made by Rosenbek *et al.* (1977) who treated a patient with anomic dysphasia with a time series design with 8 alternating one month periods of intensive treatment and no treatment. The study began more than six months after onset. During the first two treatment periods the patient's anomia resisted treatment, although over-all performance on the Porch Index of Communicative Ability (Porch, 1971) improved significantly. Suddenly 5 months after the study began rapid changes occurred in confrontation naming and spontaneous speech within a four week period. This sudden spurt of recovery is reminiscent of Wepman's patient. Rosenbek *et al.* (1977) offer three possible interpretations for this phenomenon. 1. Sudden improvement after a protracted period of treatment may be analogous to summation. In summation a neuron must be acted upon by several other neurons before it fires, so the ability to name may have resulted from the cumulative effect of several hundred stimulations from the clinician. 2. It may also be that naming performance was improving gradually over time, but that the scoring system failed to measure these gradual changes. 3. Finally the authors pointed out that the unusual recovery pattern may have been influenced by the etiology (closed head trauma), the handedness (left handed), or the age (56 yrs.) of the patient. These latter factors created questions in the authors' minds as to whether treatment had any significant influence on the patient's recovery, even though more progress was made during treatment than during non-treatment periods. They also questioned whether their approach to treatment which began with oral spelling and writing as cues to facilitate naming of common nouns was any more effective than some other treatment. As was mentioned in the beginning sections of this chapter, the influence of natural recovery, etiology, and differential treatments must always be weighed before claims are made from specific treatments. Rosenbek *et al.* (1977) remind us of this fact.

*Re-Establishing Access to the Names of Things*

In contrast to the view of anomia as a reduction in the lexical store, anomia may be viewed as an underlying loss of efficiency in retrieving words from the lexical store (Wiegel-Crump and Koenigsknecht, 1973). These investigators undertook a treatment study in an effort to investigate whether anomia represents an underlying loss of efficiency in retrieving words or actual reduction of the lexical store itself. To this end they drilled 4 anomic dysphasics on 20 of 40 initially failed words. For 18 sessions each of the 20 words was presented singly, in simple sentences, and with verbal, gestural, synonym, carrier phrase and initial phoneme cues. Following this, all 40 words were again tested. Significant improvement was noted on both the 20 drilled and 20 non-drilled words outside the superordinate category of drilled words. The authors interpreted these data as support for the argument that anomia represents impaired ability to retrieve pieces of information from the lexical store.

Oldfield (1966) also felt that anomia represented a faulty retrieval mechanism rather than reduction of the lexical store itself. He based this view at least in part upon the observation that anomic patients may show that they know what the object is by referring to the function of objects during circumlocutory speech. It has also been observed that some anomic patients resort to gestures as a kind of self cueing technique. For example, Weisenberg and McBride (1935) reported that one patient was able to say "scissors" after wiggling his fingers back and forth as in cutting. Along possibly similar lines Johnson and Rubens (1975) reported that a patient with visual agnosia was able to name real objects only when requested to remain silent and actually demonstrate the objects' use.

It has been our observation that while some patients resort to gestures which may or may not be successful in triggering off successful word retrieval, other patients make no attempt to use appropriate spontaneous gestures when struggling to name items. A critical question from the therapeutic point of view is what relationship exists between the use of gestural cues and successful word retrieval.

In 1976 a pilot study was undertaken to examine the above issue more closely (Helm, Kaplan, Vercrusze, unpublished study). A total of 17 fluent and non-fluent dysphasic patients who ranged in severity were studied under two conditions. In the first condition subjects were asked to name 24 pictures, twelve of which were classified as operative insofar as they represented objects which are easily manipulated, for example, "razor"; and twelve of which were classified as figurative insofar as they represented objects which are not easily operated upon, for example "tree". Within the two classes of pictures, 6 were highly familiar, for example "razor" and "tree" and 6 were less familiar for example "compass (drawing)" and "trellis". Not only were all verbal responses recorded and classified, but all spontaneous gestures which took place during attempts to name were noted and classified. In a second condition subjects were shown the same 24 pictures, but this time they were asked not to name the pictures, but to "show me about the picture using your hands". All gestural and spontaneous verbal responses were recorded and classified.

A significant relationship was found between the ability to name in the first condition and the ability to use appropriate gestures in the second condition, that is, the patients' with better naming scores (Condition I) demonstrated a richer gestural system (Condition II). It was also found that some patients successfully cued themselves with gestures during attempts to name, while other patients made few, random, or unrefined gestures and were not successful in retrieving names.

These preliminary data appear to support the hypothesis that anomia may result from the failure to "recall" the appropriate sensory-motor schemata which are associated with some objects and give rise to the names of these objects.

There is some anecdotal support for this hypothesis in the reports of dysphasic patients who have spontaneously "named" items when learning a non-verbal gesture system which requires them to pantomime. For example, patients who are being taught the gesture for drinking may spontaneously produce the word "cup". These informal observations and the findings of the above pilot study have encouraged Helm (unpublished data) to explore a new approach to treatment of anomia. This involves introducing the patients to a series of real objects which may be manipulated in distinct ways, for example, a hammer. Using 10 objects at a time the patient is encouraged first to use the objects and then to pantomime their use. They are also encouraged to use verbal description of function during the name retrieval process. For some anomic patients this approach has been distinctly successful insofar as they not only begin naming the treatment objects, but they improved in their ability to name objects in standardized naming tests.

This same gestural approach is being explored in picture naming tests, with equally encouraging results. For example, one patient who had moderately severe naming problems following a thalamic hemorrhage was being given the Boston Naming Test (Kaplan, Goodglass, Weintraub, 1978), and failed to name many items, lapsing instead into English and neologistic jargon. The following interchange occurred during attempts to name telescope.

Examiner: "What is this?" (showing line drawing of telescope).

Patient: "A spine-tractible with metal held on in the front and rear and projects to make something objectable to your eyes."

Examiner: "Show me about it with your hands."

Patient: (Diffuse hand wavings) "a small glass here" (pointing) "and a big glass there" (pointing) "and idagonizes for me".

Examiner: "Do this" (making a circle with thumb and forefinger and bringing to one eye, closing the other).

Patient: (imitating examiner) "a peepsader" or a "*telescope*".

# 6. Treatment of Comprehension Disorders

Many dysphasics have some degree or variety of auditory comprehension disorder. Schuell (1953) studied 130 patients having various types of dysphasia and found that all made errors on tests for comprehension of spoken language. Luria (1966) points out, however, that although comprehension disorders are common to dysphasics, the underlying mechanism for the disorder probably varies among syndromes. For example, he feels that the comprehension disorder which occurs in motor dysphasia represents a loss of verbal thought and inner speech, while the comprehension disorder in semantic dysphasia represents an inability to grasp entire grammatical structures.

Although auditory comprehension disorders may be a part of many dysphasic syndromes, it is *sensory dysphasia* that is most commonly associated a deblocking technique to reintegrate the mechanism for correlating sound and with the inability to understand the spoken word (Mills, 1904; Froschels, 1933; Nielson, 1948). Hecaen and his colleagues (Hecaen and Albert, 1978) describe 3 types of sensory dysphasia, which may represent disorders of 1. phonemic decoding 2. semantic comprehension 3. nonlinguistic attention. Because Wernicke was the first to describe sensory dysphasia, as the disorder in which understanding of speech is lost and reading and writing are severely impaired, while hearing and articulation are intact, this complex of symptoms is also known as *Wernicke's Dysphasia*. In this section the terms sensory and Wernicke's dysphasia will be used interchangeably. Unfortunately, much less has been written about treatment of severe disorders of auditory comprehension than about severe disorders of verbal production, although each disorder represents a devastating handicap.

## Reteaching Comprehension of the Spoken Word

Sensory dysphasia may represent the loss of knowledge about the meaning of spoken words. This model is suggested by the methods which attempt to reteach auditory comprehension to adult dysphasics from the simplest sound units to the most complex messages. This approach is analogous to the sound building approach described in the section on rehabilitating severely non-fluent speech. The major difference is that whereas in the case of non-fluent speech the clinician seeks to teach the patient to *produce* simple sounds, then words, then connected speech, in the case of defective auditory comprehension he instead seeks to teach the sensory dysphasic to *understand* what he hears in this same sequence. For example, Froschels (1933), who was cited in the section on severe non-fluent speech as a proponent of the sound building approach, stated that the sensory dysphasic must relearn sounds and words through the ear. Instead of sitting in front of the patient or beside him in front of a mirror, however, the clinician should stand behind the patient and have him repeat sounds, then words and then connected speech.

Huber (1944) did an extensive analysis of the verbal repetition of one Wernicke dysphasic. She discovered that a hierarchy of difficulty existed when the patient was asked to repeat vowels, diphthongs, consonants, single syllable words, and bisyllabic words. She suggested that the clinician should take this hierarchy into consideration when training Wernicke's dysphasics.

Since reading comprehension is often somewhat less impaired than auditory comprehension, sensory dysphasics are often approached through the written modality. Neilson *et al.* (1948) stated that if the sensory dysphasic was taught to read, write and do arithmetic, he would automatically relearn spoken language.

## Re-Establishing Access to Comprehension of the Spoken Word

Any clinician who uses a "deblocking" technique implies that he feels that tacit knowledge is intact but that access to it has been blocked. Two such investigators are Ulatowska and Richardson (1974). These investigators used meaning in one sensory dysphasic patient who failed to show major improvement with "traditional techniques". To accomplish this, printed words were used to reinforce and provide stable representation of speech units. The patient was trained in 3 main tasks which involved repeating words, reading the words alone, and sequencing the words in sentences. This training was given over a one year period, and the patient was observed to progress from jargon dysphasia, through phonemic and verbal paraphasias to grammatical category substitutions and finally semantic substitution.

During the ongoing Melodic Intonation Therapy study at the Boston Veterans Administration Hospital, we investigated the use of this singing technique with one severe Wernicke's dysphasic. The patient was able to reproduce the melody and rhythm patterns with ease, but was unable to reproduce the correct words of the target phrase. Instead he filled in the rhythm pattern with his own neologisms and paraphasias. Furthermore, despite the intense stimulation of about 200 verbally presented phrase-stimuli repeated over a three month period of daily treatment, neither the patient's speech nor auditory comprehension post test scores showed improvement. Unlike the method described by Ulatowska and Richardson, Melodic Intonation Therapy does not employ written stimuli, so that the patient must rely on the verbally presented material, which provides a less stable representation.

The reader is reminded that the rhythm and melody patterns used in a Melodic Intonation Therapy are based upon that of the *spoken phrase* and bear no resemblance to well known "tunes". Interestingly enough, recent work by Helm (unpublished data) with a Wernicke's dysphasic has shown that improved comprehension occurred when certain words and phrases were presented within the context of popular songs. For example, the patient was shown maps of the United States and the world, and asked to point out certain cities, states and countries. He was able to comprehend and locate six of the 12 spoken place names. The remaining six seemed to make no sense to him, even though he repeated them correctly. For example, he repeated "Chicago Chicago" and searched around California, a perseverative response from having just pointed to San Francisco correctly. After some minutes the same stimulus was sung to the familiar song "Chicago, Chicago, that totterin' town". The patient quickly said "Oh, Chicago!" and pointed it out correctly. This same phenomenon was repeated for the other five failed place names. Whereas he did not seem to understand the spoken words, he immediately recognized them within popular songs. This same patient had difficulty reading. His oral reading was paraphasic and his reading comprehension was poor. If a phrase like "For me and my gal" was given to him in written form, he would be unable to read it aloud as written and admitted that it made no sense to him. If the clinician then began singing the song from which the phrase was drawn, he immediately recognized it as evidenced by the oral phrase production. The patient stated that the phrases didn't mean anything until he heard them in a familiar song. The recognition, when it occurred via this method, was so quick, one had little doubt that the patient had suddenly gained access to what he knew about language. This result stands in hopeful contrast to the often futile attempts one makes in helping such a patient understand a particular spoken or written word, and deserves further investigation.

# C General Considerations for Dysphasia Rehabilitation

No matter which theoretical viewpoint of dysphasia the language clinician holds, or which resultant approach to treatment he chooses, there are some general considerations for improving the overall effectiveness of the rehabilitative process. In this section we will discuss some of these considerations.

## 1. Group versus Individual Language Therapy

The use of group treatment in dysphasia rehabilitation began in earnest during World War II when large numbers of dysphasic veterans were brought together in special hospital units. Although these veterans usually received individual language treatment, they also attended group sessions for both speech and psycho-sociotherapy (Amos, 1948). Since that time there has been some controversy regarding the effectiveness of group therapy for dysphasic patients.

The recently completed Veterans Administration Cooperative Treatment Study compared the progress of patients receiving non-specific group language treatment with those receiving specific individual treatment (Wertz, 1978). Patients were rated on various standardized tests before and after treatment. Those receiving individually administered therapies had significantly greater post-treatment gains in overall Porch Index of Communicative Abilities (PICA) scores than those enrolled in the groups. Since this report was not published at the time we wrote this book, we had no information as to the exact nature of the treatments which were administered. Furthermore, specific information as to which language modalities responded to treatment will be of considerably more value than overall PICA scores.

A rather unique approach to group speech rehabilitation is described by Backus and Dunn (1947) who brought together patients with a variety of disorders, such as stutterers, voice disorders, and dysphasics. They illustrated the effectiveness of this approach through several case studies. The general group approach adopted by Backus contrasts with specific group language treatment approaches such as that described by Corbin (1951), who brought together veterans with motor dysphasias and dysarthrias of varying etiologies for 30—40 minutes each day for specific treatment, which followed the motor placement, sound building approach described in the section on severe non-fluent dysphasia.

We know of no formal study which has compared the effects of specific treatments administered to groups with the effects of the same treatment administered individually. Such a study has obvious and important clinical implications.

*Psycho-Social Group Therapy*

There is little doubt that dysphasia presents psychological and sociological as well as language problems for those individuals so affected. The dysphasic patient often must deal with multiple role changes in his personal and voca-tional life. He often feels frustrated, sorry for himself, and angry. These re-actions can and do influence the dysphasic patient's response to language rehabilitation. For this reason, it is appropriate to include some mention of psycho-social group treatment in a chapter on dysphasia rehabilitation, al-though there is little literature to be examined on this subject compared to that relating to psycho-social group therapy for other disorders such as stuttering. This discrepancy is perhaps understandable when one considers that verbal exchange is an important part of the psycho-social approach and dysphasics have difficulty either talking or listening or both. Nevertheless, a few clinicians have felt that dysphasics have a need for psycho-social therapy and so have instituted such a group with their patients (Blackman and Tureen, 1948; Blackman, 1950; Aronson, Shatin, and Cook, 1956; Schlan-ger and Schlanger, 1970). The last study used pantomime as a means of circumventing the verbal communication problems of dysphasia, although it was observed that increased verbalization often accompanied the pantomimic situations.

Within the past year a psycho-social therapy group for dysphasics was begun at the Boston Veterans Medical Center. The original group leaders included the unit social worker, a speech pathologist, and a clinical psychol-ogist who himself had become dysphasic. Meetings are held twice weekly for one and a half hours. During the early sessions the group presented more problems for the professionals than for the dysphasics who attended willingly and readily. The reason for this is that many of the approaches used in this form of therapy because they are highly effective with non-dysphasic adults are not effective with dysphasics. For example, silence is often a useful thera-peutic tool with fully verbal adults but for dysphasics, silence can be devastating and totally counter-productive. Thus, the principles and ap-proaches of group work must be modified for use with dysphasic patients in psycho-social therapy groups (Blair, Walsh, and Cerny, 1979).

## 2. The Influence of Fatigue and Perseveration on Dysphasia Rehabilitation

Kurt Goldstein (1944) discussed various physiological aspects of recovery from central nervous system disorder. Two aspects which are particularly relevant to dysphasia rehabilitation are fatigue and perseveration.

### Fatigue

Goldstein stated that fatigue is due to the consumption of energy by work and can be measured according to increasing impairment in performing continuous tasks. This impairment is manifested not merely by a slowing down but by fluctuation involving discomfort, uncertainty and distress.

Two formal studies examined the effects of fatigue (Marshall, King, and Philips, 1973) and relaxation (Marshall and Watts, 1976) on the communicative ability of dysphasic adults. The former study compared standardized test scores earned by patients before and after 30 minutes of rest or 30 minutes of exercise with a Cybex electromechanical machine. Fourteen of sixteen patients had lower overall PICA scores following exercise. The latter study compared the effects of 30 minutes of relaxation exercises such as contracting and relaxing muscle groups with 30 minutes of uninterrupted quiet. The results indicate that dysphasic verbal communication is positively influenced by relaxation by virtue of the higher overall test scores earned by the treatment group immediately following relaxation exercises. Presumably dysphasic patients would respond better to the language rehabilitation process if sessions were scheduled after a rest period and included initial relaxation exercises.

### Perseveration

Goldstein (1944) stated that perseveration, a frequent secondary phenomenon in brain damage, is a way of avoiding a catastrophic response. It occurs when the patient is forced to fulfill tasks which are beyond his capabilities. If the patient is given a task which he can handle, the perseveration disappears.

In order to test this hypothesis Helm (unpublished data) experimented with one dysphasic patient whose primary response to all verbal subtests of the Boston Diagnostic Aphasia Examination (Goodglass and Kaplan, 1972) was "not, not". This verbal perseveration also dominated his daily communicative attempts, although he had been heard to use some spontaneous, overlearned expressions such as "hi" and "oh boy". It was found that whereas he responded with "not, not" to all the words in the BDAE repetition subtests, he was able to repeat "hi" and "oh boy" without perseveration. Through trial and error over 50 words were discovered to be within the patients' repetition capacity. When this list was presented to him he could repeat the entire list without perseveration. If a word which he had failed to repeat correctly in the past, for example. a word from the BDAE subtest, was inserted in the list, he produced either his last correct response or reverted to the stereotypy "not, not" for that one difficult word.

Perseveration often plagues attempts to treat patients with Melodic Intonation Therapy. One way of avoiding this is to alter the melody and rhythm of consecutive items. Recent experimentation with one severely non-fluent patient, however, showed that if the phonetic makeup and length of the phrase items were within the patient's verbal potential, the same melodic and rhythmic pattern could be repeated as many as twenty times with no perseveration. If an item which was either too long or too phonetically complex was inserted in the lesson, then the patient perseverated on the previous item.

These experiments indicate that perseveration can be controlled during the therapeutic process. Furthermore, both patients described above expanded their verbal capacities with these techniques so that items which had been too difficult and therefore produced perseveration earlier, were handled successfully at a later date.

The above techniques are reminiscent of what Muma (1977) called "systematic extension of the available repetoire", an approach to language disordered children. A series of exploratory operations are performed in order to discover the range and variations of the child's verbal capacities. The result of this exploration is that his verbal repertoire extends.

In 1962 Goda recommended that clinicians monitor the speech attempts of dysarthric patients in order to establish base line materials of therapy. In this manner he was able to derive a 26 word vocabulary for use with a dysarthric patient. New words were chosen according to the sounds produced in the spontaneous words. Finally these special words were combined into paragraphs. As the title of his paper implied Goda considered spontaneous speech to be a primary source of therapy materials.

One patient in our service (Helm, unpublished data) was recently enrolled in a program dedicated to a systematic exploration and extension of his available verbal repetoire. The patient was diagnosed by several neurologists as having a form of severe conduction dysphasia of two months duration. His speech was characterized by a great number of literal and phonemic paraphasias of which he was aware. This caused him to inhibit his own verbal output. His auditory comprehension was good but repetition was poor, as was oral reading. The clinician began treatment by writing words on the blackboard which the patient had been heard to utter in spontaneous conversation. If the patient could read the word correctly with no hesitation, it was placed on a $3 \times 5$ card for self practice. During the first session he was able to read only five words aloud. By the twelfth session he had a list of over one hundred target words he could read aloud. At that point various target words were combined to form phrases. For example "good" and "boy" became "good boy". At some point during this process the patients' spontaneous speech improved, so that, for example, he was able to tell the nurse "I want to go home" without paraphasias. Unfortunately social circumstances necessitated discharge, so that only a month was given to exploration of this technique. It would appear worthy of further investigation, however.

## 3. The Influence of Reinforcement and Feedback on Dysphasia Rehabilitation

Many of the therapeutic approaches currently used with dysphasics employ differential positive reinforcement. This means that correct responses are randomly reinforced, usually verbally. While this technique has been found to be successful with normals who are learning a task, differential positive reinforcement may make no difference in the performance of patients with dysphasia (Brookshire, 1978). Negative instructions do seem to make a difference, however (Stoicheff, 1960). This clinician gave three types of motivating instructions to 42 dysphasic subjects. The subjects who were given discouraging instructions did significantly poorer on language tasks than those given encouraging or non-evaluative instruction. Like Brookshire, Stoicheff found no difference between encouraging and non-evaluative instructions.

Chester and Egolf (1974) found that many people, clinicians included, give non-verbal negative messages to dysphasic patients. They often fail to wait until the dysphasic is finished speaking; they react to the dysphasic as if he were a non-person, speaking in the third person and walking past him without speaking; and they speak in unnecessarily loud voices. These authors concluded that many people assume that when the individuals' verbal ability is impaired, all channels of communication are impaired.

Goodkin (1969) undertook to modify the verbal behavior of spouses of three long term dysphasic patients and to evaluate the effects on the dysphasics' speech. Recordings were made of conversation between husband and wife at home before and after training. During the 15 training sessions the non-dysphasic spouse was given instruction for verbal interaction with the dysphasic spouse through earphones. These instructions followed the principles of operant conditioning. The non-dysphasic spouses learned to use positive feedback while inhibiting negative feedback and general output. At the same time, the dysphasic patient's general output increased.

## 4. The Influence of Repetition on Dysphasia Rehabilitation

In the early sections of this chapter we mentioned that some therapeutic approaches to dysphasia rely heavily on drills in which a particular sound, word or phrase is repeated. Even treatments which are less drill oriented, such as Melodic Intonation Therapy, require multiple repetitions of each target phrase. Little research attention seems to have been given to the well known fact that words may lose their meaning after multiple repetitions. This phenomenon is called verbal satiation.

In 1919 Bassett and Warne studied the lapse of verbal meaning with repetition. These investigators sought to determine how many repetitions were required before everyday monosyllabic nouns lost their meaning. They found that the faster the repetition, the quicker the lapse of meaning. In addition, the meaning of familiar monosyllabic nouns which are repeated aloud about three times per second drop away in about three to 3.5 seconds or in

9 repetitions. This study was carried out with normal subjects. It might well be that brain damaged patients would experience verbal satiation with fewer repetitions. Miller (1963) found that the meanings of words could be maintained longer when concurrent, appropriate activity accompanied the repetition. For example, if the subject actually pushed against something while saying "push", the meaning of that word was retained significantly longer than when the word was repeated without such activity, or with an action not usually paired with it. Although Miller's subjects were normals, this study appears to have direct applicability to dysphasia rehabilitation. Along these lines, West (1977) reviewed the literature on imagery and concluded that *action* imagery is more memorable than *static* imagery and the clinician would do well to employ pictures of an object being manipulated rather than simple line drawings of the object alone.

One study examined the effects of repetition on naming behavior in one anomic dysphasic (Helmick and Wipplinger, 1975). Forty-five words were presented to the patient for pre-treatment naming. The list was divided into three lists of 15 each, which received 24,6 and no repetitions respectively per treatment session. The post-test results eight weeks later led the authors to conclude that the number of repetitions did not affect the patient's naming behavior, but training in general did affect the patient's performance.

# 5. The Influence of Sensory Input Variables on Dysphasia Rehabilitation

Clinicians have long been advised to speak slowly to dysphasic patients, as this is felt to increase auditory comprehension (Mills, 1904; Schuell, Jenkins, Jiménez-Pabón, 1964). Albert and Bear (1974) investigated the effect of abnormally slow rate of presentation on the understanding of a man with word deafness. They found that the patient's comprehension was significantly better for digit trigrams presented with 3 second separations, than for the same trigrams presented with no separation. Gardner, Albert, and Weintraub (1975) undertook to assess the effects of speed on the auditory word comprehension of a series of dysphasic patients. At the same time they examined the effects of redundancy on auditory word comprehension. This latter condition involved embedding the target word in a sentence which contained a semantically supportive word. For example, a low redundancy or neutral sentence would be "The grass is nice" a high redundancy sentence, "The grass is green". Target words were presented alone, in low redundancy sentences presented at normal and slow speeds, and in high redundancy sentences presented at normal and slow speeds. Both sentence redundancy and rate of presentation made a contribution to comprehension. The authors suggested that treatment procedures should present target words in slowly enunciated semantically redundant utterances, then gradually increase the rate of speaking and eliminate the redundant cues.

Weidner and Lasky (1976) compared the effects of normal oral presentation rate (150 words per minute) and slowed rate (110 words per minute) on the performance of 20 dysphasic patients. The tasks required patients to identify pictures, answer "yes-no" questions, follow commands, repeat sentences. The results indicate the patients make fewer errors with the slower rate, especially with identifying pictures and following commands.

McDearmon and Potter (1975) reviewed various studies which looked at the influence of various sensory modalities (for example, visual, tactile, auditory, olfactory) on the overall naming performance of dysphasics. They conclude that while most dysphasics tend to show similarity of overall naming disability across modalities, some do better with a particular modality and some severe cases do better with stimulus redundancy. Clinicians therefore should be prepared to explore multimodal and multisensory cues with dysphasic patients.

## 6. Conclusion

Dysphasia rehabilitation has a rich history representing nearly a century of tireless investigation and application of remedial approaches. During this time speech pathologists, neurologists and psychologists alike have exerted considerable clinical effort towards helping the dysphasic patient to overcome his language handicap. As we have seen through examination of detailed case studies and large group studies, this effort has made a significant difference in the level of communication skills ultimately regained by many dysphasic patients. As we have also seen, there is very little which is entirely new in dysphasia rehabilitation. Instead, we find that most of our present techniques have evolved from earlier observations and methods, which have been enriched with knowledge from emerging fields of learning psychology, neurolinguistics, neuropsychology and statistics. Perhaps most importantly, we have seen that through careful differential diagnosis of adult dysphasia, it may be possible to develop successful differential treatments for adult dysphasics.

# References

Aimard, G., Devic, M., Lebel, M., Trovillas, P., Boisson, D. (1975): Agraphie (Dynamique?) d'origine frontal. Rev. Neurol. *131*, 505—512.

Ajax, E. T. (1967): Dyslexia without dysgraphia. Arch. Neurol. *17*, 645—651.

Ajax, E. T. (1977): Alexia without agraphia and the inferior splenium. Neurol. *27*, 685—688.

Alajouanine, T., et al. (1957): Étude de quarante-trois cas d'aphasie posttraumatique. L'Encéphale *161*, 1—45.

Alajouanine, T., Castaigne, P., Sabourand, O., Contamin, F. (1959): Palilalie paroxystique et vocalisations itératives au cours de crises épileptiques, etc. Rev. Neurol. *101*, 685—697.

Alajouanine, T., Lhermitte, F. (1964): Les composantes phonémiques et sémantiques de la jargonaphasie. Int. J. Neurol. *4*, 277—286. Reprinted in: Psycholinguistics and Aphasia (Goodglass, H., Blumstein, S. E., eds. and trans.), pp. 319—329. Baltimore: The Johns Hopkins University Press. 1973.

Alajouanine, T., Lhermitte, F. (1973): The Phonemic and Semantic Component of Jargon Aphasia. In: Psycholinguistics and Aphasia (Goodglass, H., Blumstein, S. E., eds.), pp. 318 to 329. Baltimore: The Johns Hopkins University Press.

Alajouanine, T., Lhermitte, F., Ledoux, D., Renaud, D., Vignolo, A. (1964): Les composantes phonémiques et sémantiques de la jargonaphasie. Rev. Neurol. *110*, 5—20.

Alajouanine, T., Ombredane, A., Durand, M. (1939): Le syndrome de désintégration phonétique dans l'aphasie. Paris: Masson. 1939.

Alajouanine, T., Pichot, P., Durand, M. (1949): Dissociation des altérations phonétiques avec conservation relative de la langue la plus ancienne dans un cas d'anarthrie pure chez un sujet français bilingue. L'Encéphale *38*, 245—265.

Alajouanine, T., Sabourand, O., De Ribaucourt, B. (1952): Le jargon des aphasiques. J. Psychol. *45*, 158—180, 293—329.

Albert, M. L., Bear, D. (1974): Time to understand. A case study of word deafness with reference to the role of time in auditory comprehension. Brain *97*, 383—394.

Albert, M. L., Obler, L. K. (1975): Mixed polyglot aphasia. Paper presented at 13th Annual Meeting, Academy of Aphasia, Victoria, British Columbia, October 1975.

Albert, M. L., Obler, L. K. (1978): The Bilingual Brain: Neurolinguistic and Neuropsychological Aspects of Bilingualism. New York: Academic Press.

Albert, M. L., Sparks, R., Helm, N. (1973): Melodic intonation therapy for aphasia. Arch. Neurol. *29*, 130—131.

Albert, M. L., Sparks, R., von Stockert, T., Sax, D. (1972): A case of auditory agnosia: linguistic and nonlinguistic processing. Cortex *8*, 427—443.

Albert, M. L., Yamadori, A., Gardner, H., Howes, D. (1973): Comprehension in alexia. Brain *96*, 317—328.

Alexander, M. P., LoVerme, S. R. (1980): Aphasia following left hemispheric intracerebral hemorrhage. Neurol. *30*, 1193—1202.

Alexander, M. P., Schmitt, M. A. (1980): The aphasia syndrome of stroke in the left anterior cerebral artery territory. Arch. Neurol. *37*, 97—100.

Amos, M. L. (1948): Speech rehabilitation for veterans. Quart. J. Sp. *34*, 73—79.

Andreewsky, E., Seron, X. (1975): Implicit processing of grammatical rules in a classical case of agrammatism. Cortex *11*, 379—390.

Angelergues, R., Hecaen, H., Djindjian, R., *et al.* (1962): Un cas d'aphasie croisée. Rev. Neurol. *107*, 543—545.

Arnaud, R. (1887): Contribution à l'étude clinique de la surdite-verbale. Arch. de Neurol. *13*, 177—200.

Aronson, M., Shatin, L., Cook, J. (1956): Sociopsychotherapeutic approach to the treatment of aphasic patients. JSHD *21*, 352—364.

Arseni, C., Botez, M. I. (1961): Speech disturbances caused by tumors of the supplementary motor area. Acta Psychiat. Scand. *36*, 279—299.

Assal, G., Chapuis, G., Zander, E. (1970): Isolated writing disorders in a patient with stenosis of the left internal carotid artery. Cortex *6*, 241—248.

Backus, O. L. (1937): The Rehabilitation of Persons with Aphasia. In: The Pathology of Speech, Chap. 25, pp. 439—466. New York-London: Harper and Bros., Publisher.

Backus, O. L. (1945): The rehabilitation of aphasic veterans. JSHD *10*, 149—153.

Backus, O. L., Dunn, H. M. (1947): Intensive group therapy in speech rehabilitation. JSHD *12*, 39—60.

Baker, E., Berry, T., Gardner, H., Zurif, E., Davis, L., Veroff, A. (1975): Can linguistic function be dissociated from natural language? Nature *254*, 509—510.

Barraquer-Bordas, C., Mendilaharsu, C., Peres-Serra, J., *et al.* (1963): Estudio de dos casos de afazia cruzada en pacientes manidextros. Acta Neurol. Latinamer. *9*, 140—148.

Barten, M. I., Maruszewski, M., Urrea, D. (1969): Variation of stimulus context and its effect on word-finding in aphasia. Cortex *5*, 351—365.

Bassett, M. F., Warne, C. J. (1919): On the lapse of verbal meaning with repetition. Am. J. Psychol. *30*, 415—418.

Basso, A., Capitani, E., Vignolo, L. A. (1979): Influence of rehabilitation on language skills in aphasia patients. Arch. Neurol. *36*, 190—195.

Basso, A., Faglioni, P., Vignolo, L. A. (1974): Étude controlée de la rééducation du langage dans l'aphasie: comparison entre aphasiques traités et non-traités. Revue Neurologique *131*, 607—614.

Basso, A., Taborrelli, A., Vignolo, L. A. (1978): Dissociated disorders of speaking and writing in aphasia. J. Neurol. Neurosurg. Psychiat. *41*, 556—563.

Bastian, H. C. (1887): On different kinds of aphasia. Brit. Med. J. *2*, 931—937, 985—990.

Bastian, H. (1897): Some problems in connexion with aphasia and other speech defects. Lancet *1*, 933—942, 1005—1017, 1131—1137, 1187—1194.

Bell, D. S. (1968): Speech functions of the thalamus inferred from the effects of thalamotomy. Brain *91*, 619—638.

Benson, D. F. (1967): Fluency in aphasia: Correlation with radioactive scan localization. Cortex *3*, 373—392.

Benson, D. F. (1973): Psychiatric aspects of aphasia. Brit. J. Psychiat. *123*, 555—566.

Benson, D. F. (1977): The third alexia. Arch. Neurol. *34*, 317—322.

Benson, D. F. (1978): Amnesia. South. Med. J. *71*, 1221—1228.

Benson, D. F. (1979): Aphasia rehabilitation. Arch. Neurol. *36*, 187—189.

Benson, D. F. (1979): Neurologic Correlates of Anomia. In: Studies in Neurolinguistics, Vol. 4, pp. 293—328. New York: Academic Press.

Benson, D. F., Brown, J., Tomlinson, E. (1971): Varieties of alexia. Neurol. *21*, 951—957.

Benson, D. F., Geschwind, N. (1969): The Alexias. In: Handbook of Clinical Neurology (Vinken, P., Bruyn, G., eds.). Amsterdam: North-Holland.

Benson, D. F., Geschwind, N. (1976): The Aphasias and Related Disturbances. In: Clinical Neurology (Baker, A. B., Baker, L. H., eds.). Hagerstown: Harper and Row.

Benson, D. F., Marsden, C. D., Meadows, J. C. (1974): The amnesic syndrome of posterior cerebral artery occlusion. Acta Neurol. Scand. *50*, 133—145.

Benson, D. F., Segarra, J., Albert, M. L. (1974): Visual agnosia-prosopagnosia, a clinico-pathological correlation. Arch. Neurol. *30*, 307—310.

Benson, D. F., Sherematta, W. A., Bouchard, R., Segarra, J. M., Price, D., Geschwind, N. (1973): Conduction aphasia: A clinicopathological study. Arch. Neurol. *28*, 339—346.

Benson, D. F., Tomlinson, E. B. (1971): Hemiplegic syndrome of the posterior cerebral artery. Stroke 2, 559—564.

Benton, A. L. (1945): A visual retention test for clinical use. Arch. Neurol. Psychiat. *54*, 212—216.

Benton, A. L. (1950): A multiple choice type of the visual retention test. Arch. Neurol. Psychiat. *64*, 699—707.

Benton, A. L. (1968): Differential behavioral effects in frontal lobe disease. Neuropsychologia *6*, 53—60.

Benton, A. L., Joynt, R. J. (1960): Early descriptions of aphasia. Arch. Neurol. *3*, 205—222.

Beyn, E. S., Shokhor-Trotskaya, M. K. (1966): The preventive method of speech rehabilitation in aphasia. Cortex 2, 96—108.

Biemond, A. (1956): The conduction of pain above the level of the thalamus opticus. Arch. Neurol. Psychiat. *75*, 231—244.

Blackman, N. (1950): Group psychotherapy with aphasics. J. Nerv. Men. Disease *111*, 154—163.

Blackman, N., Tureen, L. (1948): Aphasia: Psychosomatic approach in rehabilitation. Trans. Amer. Neurol. Assoc. *73*, 193—196.

Blair, J., Walsh, M., Cerny, S. (1979): The group as an adjustment modality for aphasia patients. Presented to American Society for Group Psychotherapy and Psychodrama Annual Meeting, New York, April 1979.

Blumstein, S. E. (1973): A Phonological Investigation of Aphasic Speech. The Hague: Mouton.

Blumstein, S. E., Baker, E., Goodglass, H. (1977): Phonological factors in auditory comprehension in aphasia. Neuropsychologia *15*, 19—30.

Blumstein, S. E., Cooper, W. (1974): Hemispheric processing of intonational contours. Cortex *10*, 146—158.

Bogen, J. E. (1976): Linguistic Performance in the Short-Term Following Cerebral Commissurotomy. In: Studies in Neurolinguistics, Vol. 2 (Whitaker, H., Whitaker, H., eds.). New York: Academic Press.

Boller, F. (1973): Destruction of Wernicke's area without language disturbance. A fresh look at crossed aphasia. Neuropsychologia *11*, 243—246.

Boller, F., Green, E. (1972): Comprehension in severe aphasics. Cortex *8*, 382—394.

Boller, F., Vignolo, L. A. (1966): Significato dei Disturbi della Ripetizione nell'Afasia di Wernicke. Sistema Nervoso *18*, 383—396.

Bönhoeffer, K. (1914): Klinischer und anatomischer Befund zur Lehre von der Apraxie und der „motorischen Sprachbahn". Mschr. Psychiat. Neurol. *35*, 113—128.

Bönhoeffer, K. (1923): Zur Klinik und Lokalisation des Agrammatismus und der Rechts-Links-Desorientierung. Mschr. Psychiat. Neurol. *54*, 11—42.

Botez, M. I. (1962): Afazia si sindroamele corelate in procesele expansive intracraniene (Rumanian). Bucarest: Ed. Acad. Rep. Popul. Romine.

Botez, M. I., Barbeau, A. (1971): Role of subcortical structures and particularly the thalamus in the mechanisms of speech and language. Int. J. Neurol. *8*, 300—320.

Brain, W. R. (1961): Speech Disorders. London: Butterworths.

Brain, W. R. (1965): Speech Disorders, 2nd ed. London: Butterworth.

Bramwell, B. (1899): On crossed aphasia. Lancet, June 3, 803—805.

Brickner, R. (1940): A human cortical area producing repetitive phenomena when stimulated. J. Neurophysiol. *3*, 128—130.

Broadbent, W. H. (1884): On a Particular Form of Amnesia: Loss of Nouns. In: On the Cerebral Mechanisms of Speech and Thought. Med. Chir. Trans. *55*, 249—264.

Broca, P. (1865): Sur la faculté du langage articulé. Bull. Soc. Anthrop. (Paris) *6*, 337—393.

Brookshire, R. H. (1978): Recording behavioral events in aphasia treatment. Paper presented as part of symposium "A Language Rehabilitation in Aphasia". Amer. Acad. Advan. Sci. Annual Meeting, Washington, D.C.

Brown, J. W. (1972): Aphasia, Apraxia and Agnosia. Springfield, Ill.: Charles C Thomas.

Brown, J. W. (1977): Mind, Brain and Consciousness: The Neuropsychology of Cognition. New York: Academic Press.

Brown, J. W., Hecaen, H. (1976): Lateralization and language representation. Neurol. 26, 183—189.

Brown, J. W., Jaffe, J. (1975): Hypothesis on cerebral dominance. Neuropsychologia 13, 107—110.

Brown, J. W., Wilson, F. (1973): Crossed aphasia in a dextral. Neurol. 23, 907—911.

Brown, R. (1973): A First Language. Cambridge: Harvard University Press.

Brust, J., Shafer, S., Richter, R., Bruun, B. (1976): Aphasia in acute stroke. Stroke 7, 167—174.

Bugiani, O., Conforto, C., Sacco, G. (1969): Aphasia in thalamic hemorrhage. Lancet 1052.

Burns, M. S., Canter, C. J. (1977): Phonemic behavior of aphasic patients with posterior cerebral lesions. Brain Lang. 4, 492—501.

Butfield, E., Zangwill, L. L. (1946): Reeducation in aphasia: a review of 70 cases. J. Neurol. Neurosurg. Psychiat. 9, 75—79.

Bychowski, Z. (1919): Über die Restitution der nach einem Schädelschuß verlorenen Umgangssprache bei einem Polyglotten. Mschr. Psychol. Neurol. 45, 183—201.

Caplan, L. R. (1978): Variability of perceptual function. The sensory cortex as a "categorizer" and "deducer". Brain and Language 6, 1—13.

Caplan, L. R., Hedley-White, T. (1974): Cueing and memory dysfunction in alexia without agraphia: a case report. Brain 97, 251—262.

Cappa, S. F., Vignolo, L. A. (1979): "Transcortical" features of aphasia following left thalamic hemorrhage. Cortex 15, 121—130.

Caraceni, T. (1962): L'afasia di conduzione. Riv. Pat. Nerv. Ment. 83, 531—551.

Caramazza, A., Berndt, R. S. (1978): Semantic and syntactic processes in aphasia: a review of the literature. Psychol. Bull. 85, 898—918.

Caramazza, A., Zurif, E. B. (1976): Dissociation of algorithmic and heuristic processes in language comprehension: evidence from aphasia. Brain Lang. 3, 572—582.

Carrieri, G. (1963): Syndrome of disturbance of the left motor supplementary area associated with a parasagittal meningioma. Riv. Pat. Nerv. Ment. 84, 29—48.

Cermak, L. S., Moreines, J. (1976): Verbal retention deficits in aphasic and amnesic patients. Brain Lang. 3, 16—27.

Charlton, M. (1964): Aphasia in bilingual and polyglot patients. J. Sp. Hear. Dis. 29, 307—311.

Chavany, J. A., Rongerie, J. (1958): L'aphémie post-opératoire transitoire après lobectomie frontale gauche. Presse Méd. 66, 1191—1192.

Chédru, F., Geschwind, N. (1972): Writing disturbances in acute confusional states. Neuropsychol. 10, 343—353.

Cheek, W. R., Taveras, J. M. (1966): Thalamic tumors. J. Neurosurg. 24, 505—513.

Chen, L. C. Y. (1971): Manual communication by combined alphabet and gestures. Arch. of Phys. Med. and Rehab. 381—384.

Chesner, E. C. (1937): Aphasia I. Techniques of clinical examination. Bull. Neurol. Inst. 6, 134—144.

Chester, S. L., Egolf, D. B. (1974): Non-verbal communication and aphasia therapy. Rehab. Lit. 8, 231—233.

Chusid, J., Gutierrez-Mahoney, C., Margules-Lavergue, M. (1954): Speech disturbances in association with parasagittal frontal lesions. J. Neurosurg. 11, 193—204.

Ciemins, V. (1970): Localized thalamic hemorrhage: a cause of aphasia. Neurol. 20, 776—782.

Clarke, B., Zangwill, O. (1965): A case of "crossed aphasia" in a dextral. Neuropsychologia 3, 81—86.

Coenen, W. (1940): Klinischer und anatomischer Beitrag zur Frage der Leitungsaphasie. Arch. Psychiat. 112, 664—678.

Conrad, K. (1947): Über den Begriff der Vorgestalt und seine Bedeutung für die Hirnpathologie. Nervenarzt 18, 289—293.

Conrad, K. (1954): New problems of aphasia. Brain 77, 491—509.

Corbin, M. L. (1951): Group speech therapy for motor aphasia and dysarthria. JSHD 16, 21—34.

Corlew, M. M., Nation, J. E. (1975): Characteristics of visual stimuli and naming performance in aphasic adults. Cortex *11*, 186—191.

Critchley, M. (1930): Anterior cerebral artery and its syndromes. Brain *53*, 120—165.

Critchley, M. (1938): Aphasia in a partial deaf-mute. Brain *61*, 163—169.

Critchley, M. (1964): The neurology of psychotic speech. Brit. J. Psychiat. *110*, 353—364.

Crystal, D., Fletcher, P., Garman, M. (1976): The Grammatical Analysis of Language Disability. New York: Elsevier.

Culton, G. L. (1969): Spontaneous recovery from aphasia. JSHR *12*, 825—832.

Cumming, W. J. K., Hurwitz, L. L., Perl, N. T. (1970): A study of a patient who had alexia without agraphia. J. Neurol. Neurosurg. Psychiat. *33*, 34—39.

Dabul, B., Bollier, B. (1976): Therapeutic approaches to apraxia. JSHD *41*, 268—276.

Damasio, A. R., Kassel, N. F. (1978): Transcortical motor aphasia in relation to lesions of the supplementary motor area. Neurol. *28*, 396.

Damasio, A. R., Kassel, N. F. (1978): Transcortical motor aphasia in relation to lesions of anomia. Brain Lang. *7*, 74—85.

Darley, F. L. (1970): Language Rehabilitation. In: Behavioral Changes in Cerebrovascular Diseases (Benton, A. L., ed.), pp. 51—76. New York: Harper.

Darley, F. L. (1975): Treatment of Acquired Aphasia. In: Advances in Neurology, Vol. 7 (Friedlander, W. J., ed.). New York: Raven Press.

Darley, F. L., Aronson, A., Brown, J. (1975): Motor Speech Disorders. Philadelphia: W. B. Saunders.

Davis, G. H., Bisset, J. (1977): Employing single-subject research designs in evaluating aphasia therapy. ASHA Meeting, Chicago.

Davis, L., Foldi, N. S., Gardner, H., Zurif, E. (1978): Repetition in the transcortical aphasias. Brain Lang. *6*, 226—238.

Deal, J. I., Darley, F. I. (1972): The influence of linguistic and situational variables on phonemic accuracy in apraxia of speech. J. Sp. Hear. Res. *15*, 639—653.

Dejerine, J. (1891): Sur un cas de cécité verbale avec agraphie, suivi d'autopsie. Mém. Soc. Biol. *3*, 197—201.

Dejerine, J. (1892): Contribution à l'étude anatomoclinique et clinique des différentes variétés de cécité verbale. C.R. Soc. Biol. *4*, 61—90.

Dejerine, J. (1914): Sémiologie des affections du système nerveux. Paris: Masson.

Dejerine, J., Andre-Thomas, J. (1904): Un cas de cécité verbale avec agraphie suivi d'autopsie. Rev. Neurol. *12*, 655—664.

Delay, J., Brion, S. (1962): Les démences tardives. Paris: Masson.

Denes, G., Semenza, C. (1975): Auditory modality-specific anomia: evidence from a case of pure word deafness. Cortex *11*, 401—411.

Denny-Brown, D. (1963): The Physiological Basis of Perception and Speech. In: Problems of Dynamic Neurology (Halpern, L., ed.), pp. 30—62. Jerusalem: Hebrew University Press.

Denny-Brown, D., Chambers, R. A. (1958): The parietal lobe and behavior. Res. Publ. Assoc. Nerv. Ment. Dis. *36*, 35—117.

DeRenzi, E., Faglioni, P., Scotti, G., Spinnler, H. (1972): Impairment in associating color to form, concomitant with aphasia. Brain *95*, 293—304.

DeRenzi, E., Pieczuro, A., Vignolo, L. (1966): Oral apraxia and aphasia. Cortex *2*, 50—73.

DeRenzi, E., Vignolo, L. (1962): The token test: a sensitive test to detect receptive disturbances in aphasia. Brain *85*, 665—678.

Douglass, E., Richardson, J. (1959): Aphasia in a congenital deaf-mute. Brain *83*, 68—80.

Dreifuss, F. (1961): Observations on aphasia in a polyglot poet. Acta Psychiat. Scandinav. *36*, 91—97.

Dubois, J., Hecaen, H., Angelergues, R., de Chatelier, A., Marcie, P. (1964): Étude neuro-linguistique de l'aphasie de conduction. Neuropsychologia *2*, 9—44. Reprinted in: Linguistics and Aphasia (Goodglass, H., Blumstein, S. E., eds. and trans.), pp. 284—300. Baltimore: The Johns Hopkins University Press. 1973.

Dubois, J., Hecaen, H., Marcie, P. (1969): L'agraphie "pure". Neuropsychol. *7*, 271—286.

Eagelson, H. M., Vaughn, G. R., Knudson, A. B. (1970): Hand signals for dysphasia. Arch. of Phys. Med. and Rehab. *51*, 111—113.

Earnest, M. P., Monroe, P. A., Yarnell, P. R. (1977): Cortical deafness: demonstration of the pathologic anatomy by CT scan. Neurol. 27, 1172—1175.

Eisenson, J. (1949): Prognostic factors related to language rehabilitation in aphasic patients. JSHD 14, 262—264.

Eisenson, J. (1975): Language rehabilitation of aphasic adults: a review of some issues as to the state of the art. In: The Nervous System (Human Communication and Its Disorders, Vol. 3) (Tower, D. H., ed.). New York: Raven Press.

Erickson, T. C., Woolsey, C. N. (1951): Observations on the supplementary motor area of man. Tr. Am. Neur. Ass. 50—56.

Ettlinger, G., Jackson, C., Zangwill, O. (1955): Dysphasia following right temporal lobectomy in a right-handed man. J. Neurol. Neurosurg. Psychiat. 18, 214—217.

Exner, S. (1881): Untersuchungen über die Lokalisation der Funktionen in der Großhirnrinde des Menschen. Wien: W. Braumüller.

Fisher, C. M. (1959): The pathological and clinical aspects of thalamic hemorrhage. Trans. Amer. Neurol. Assoc. 84, 56—59.

Fisher, C. M. (1965): Pure sensory stroke involving face, arm and leg. Neurol. 15, 76—80.

Foix, C. (1928): Aphasies. In: Nouveau traité de médecine (Hoger, G. H., Widal, F., Teissier, P. J., eds.), Vol. 18. Paris: Masson.

Foix, C., Hillemand, P. (1925): Rôle vraisemblabe du splénium dans la pathogénie de l'alexie pure par lésion de la cérébrale postérieure. Bull. et Mém. Soc. Méd. Hop. Paris 41, 393—395.

Fraisse, P., Noizet, G., Flament, C. (1962): Fréquence et familiarité du vocabulaire. In: Problèmes de Psycholinguistique, pp. 157—167. Paris: P.U.F.

Freund, C. (1889): Über optische Aphasie und Seelenblindheit. Arch. Psychiat. Nervenkr. 20, 276—297, 371—416.

Froschels, E. (1933): Speech Therapy. Boston: Expression Co.

Galaburda, A. M., Kemper, T. L. (1979): Cytoarchitectonic abnormalities in developmental dyslexia: a case study. Annals Neurol. 6, 94—101.

Galaburda, A. M., LeMay, M., Kemper, T., Geschwind, N. (1978): Right-left asymmetries in the brain. Science 199, 852—856.

Gardner, H., Albert, M. L., Weintraub, S. (1975): Comprehending a word: The influence of speed and redundancy on auditory comprehension in aphasia. Cortex 11, 155—162.

Gardner, H., Denes, G., Zurif, E. (1975): Critical reading at the sentence level in aphasia. Cortex 11, 60—72.

Gardner, H., Zurif, E. (1975): Bee but not be: oral reading of single words in aphasia and alexia. Neuropsychol. 13, 170—181.

Gardner, H., Zurif, E. B., Berry, T., Baker, E. (1976): Visual communication in aphasia. Neuropsychologia 14, 275—292.

Gazzaniga, M. S., Hillyard, S. A. (1971): Language and speech capacity of the right hemisphere. Neuropsychologia 9, 273—280.

Gazzaniga, M., Sperry, R. (1967): Language after section of the cerebral commissures. Brain 90, 131—148.

Gerstmann, J. (1930): Zur Symptomatologie der Hirnläsionen im Übergangsgebiet der unteren Parietal- und mittleren Occipitalwindung. Nervenarzt 3, 691—695 (1930).

Geschwind, N. (1965): Disconnexion syndromes in animals and man. Brain 88, 237—294, 585—644.

Geschwind, N. (1965): The varieties of naming errors. Cortex 3, 97—112.

Geschwind, N. (1967): The Apraxias. In: Phenomenology of Will and Action (Straus, E. W., Griffith, R. M., eds.), pp. 91—102. Pittsburgh: Duquesne Univ. Press.

Geschwind, N. (1975): The apraxias: neural mechanisms of disorders of learned movement. Amer. Scientist 63, 188—195.

Geschwind, N., Fusillo, M. (1966): Color-naming defects in association with alexia. Arch. Neurol. 15, 137—146.

Geschwind, N., Kaplan, E. (1962): A human cerebral deconnection syndrome. Neurol. 12, 675—685.

Geschwind, N., Levitsky, W. (1968): Human brain: left-right asymmetries in temporal speech regions. Science 161, 186—187.

Geschwind, N., Quadfasel, F., Segarra, J. (1968): Isolation of the speech area. Neuropsychologia *6*, 327—340.

Glass, A. V., Gazzaniga, M. S., Premack, D. (1973): Artificial language training in aphasia. Neuropsychologia *11*, 95—103.

Gleason, J. B., Goodglass, H., Green, E., Ackerman, N., Hyde, M. K. (1975): The retrieval of syntax in Broca's aphasia. Brain Lang. *2*, 451—471.

Gloning, I., Gloning, K. (1965): Aphasien bei Polyglotten. Wien. Zschr. Nervenheilkunde *22*, 362—397.

Gloning, I., Gloning, K., Haub, G., *et al.* (1969): Comparison of verbal behavior in right-handed and non-right-handed patients with anatomically verified lesions of one hemisphere. Cortex *5*, 43—62.

Gloning, I., Gloning, K., Hoff, H. (1963): Aphasia: a Clinical Syndrome. In: Problems of Dynamic Neurology (Halpern, L., ed.). Jerusalem: Hebrew University Press.

Goda, S. (1962): Spontaneous speech, a primary source of therapy material. JSHD *27*, 190—192.

Godfrey, C. M., Douglass, E. (1959): The recovery process in aphasia. Canad. Med. Assoc. J. *80*, 618—624.

Goldstein, K. (1911): Die amnestische und die zentrale Aphasie (Leitungsaphasie). Arch. Psychiat. Neurol. *48*, 314—343.

Goldstein, K. (1915): Die transkortikalen Aphasien. Ergebn. Neurol. u. Psychiat. Jena: G. Fischer.

Goldstein, K. (1917): Die transkortikalen Aphasien. Jena: G. Fischer.

Goldstein, K. (1924): Das Wesen der amnestischen Aphasia. Schweiz. Arch. Neurol. Psychiat. *15*, 163—175.

Goldstein, K. (1942): After-Effects of Brain-Injuries in War: Their Evaluation and Treatment. New York: Grune and Stratton.

Goldstein, K. (1944): Physiological aspects of convalescence and rehabilitation following central nervous system injuries. Fed. Proc. *3*, 255—267.

Goldstein, K. (1948): Language and Language Disturbances. New York: Grune and Stratton.

Goldstein, K., Marmor, J. (1938): A case of aphasia, with special reference to the problems of repetition and word-finding. J. Neurol. Psychiat. I (New Series), 329—339.

Goldstein, M. N., Joynt, R. J., Goldblatt, D. (1971): Word blindness with intact central visual fields. A case report. Neurol. *21*, 873—876.

Goodglass, H. (1968): Studies on the grammar of aphasics. In: Developments in Applied Psycholinguistic Research (Rosenberg, S., Kaplan, J., eds.). New York: Macmillan.

Goodglass, H., Baker, E. (1976): Semantic field, naming and auditory comprehension in aphasia. Brain Lang. *3*, 359—374.

Goodglass, H., Barton, M. I., Kaplan, E. (1968): Sensory modality and object-naming in aphasia. J. Sp. Hear. Res. *3*, 257—267.

Goodglass, H., Geschwind, N. (1976): Language Disorders (Aphasia). In: Handbook of Perception (Carterette, E. C., Friedman, M. P., eds.), Vol. 7. New York: Academic Press.

Goodglass, H., Gleason, J. B., Hyde, M. R. (1970): Some dimensions of language comprehension in aphasia. J. Sp. Hear. Res. *13*, 595—606.

Goodglass, H., Kaplan, E. (1963): Disturbance of gesture and pantomime in aphasia. Brain *86*, 703—720.

Goodglass, H., Kaplan, E. (1972): Assessment of Aphasia and Related Disorders. Philadelphia: Lea and Febiger.

Goodglass, H., Kaplan, E. (1979): The Assessment of Cognitive Deficit in the Brain-Injured Patient. In: Handbook of Behavioral Neurobiology-Neuropsychology (Gazzaniga, M., ed.), pp. 3—22. New York: Plenum Press.

Goodglass, H., Kaplan, E., Weintraub, S., Ackerman, N. (1976): The tip-of-the-tongue phenomenon in aphasia. Cortex *12*, 145—153.

Goodglass, H., Klein, B., Carey, P., Jones, K. (1966): Specific semantic word categories in aphasia. Cortex *2*, 74—89.

Goodglass, H., Quadfasel, F. A. (1954): Language laterality in left-handed aphasics. Brain *77*, 521—548.

Goodglass, H., Quadfasel, F. A., Timberlake, W. H. (1964): Phrase length and the type and severity of aphasia. Cortex *1*, 133—153.

Goodkin, R. (1969): A procedure for training spouses to improve functional speech of aphasic patients. Proceedings 77th annual convention. Amer. Psych. Assoc.

Graham, F. K., Kendall, B. S. (1946): Memory for designs test. St. Louis: Washington University.

Grasset, P. (1896): Aphasie de la main droite chez un sourd-muet. Le Progrès Médical *4*, 281—282.

Green, E. (1969): Phonological and grammatical aspects of jargon in an aphasic patient: a case study. Lang. Speech *12*, 103—118.

Green, E., Howes, D. H. (1977): The Nature of Conduction Aphasia: A Study of Anatomic and Clinical Features and of Underlying Mechanisms. In: Studies in Neurolinguistics, Vol. 3, pp. 123—156. New York: Academic Press.

Greenblatt, S. H. (1973): Alexia without agraphia or hemianopsia. Brain *96*, 307—316.

Greenblatt, S. H. (1976): Subangular alexia without agraphia or hemianopia. Brain Lang. *3*, 229—245.

Guidetti, B. (1957): Disturbi della parola associati a lesioni della parte posteriore dell'area supplementare motoria. Riv. Neurol. Nap. *27*, 195—201.

Halpern, L. (1941): Beitrag zur Restitution der Aphasie bei Polyglotten im Hinblick auf das Hebräische. Schweiz. Arch. Neurol. Psychiat. *47*, 150—154.

Hatfield, F. M. (1972): Looking for help from linguistics. Brit. J. Dis. Communic. *7*, 64—81.

Hatfield, F. M., Howard, D., et al. (1977): Object naming in aphasia—the lack of effect of context or realism. Neuropsychologia *15*, 717—727.

Head, H. (1926): Aphasia and Kindred Disorders of Speech. New York: Macmillan.

Hecaen, H. (1969 a): Aphasic, Apraxic, and Agnosic Syndromes in Right and Left Hemisphere Lesions. In: Handbook of Clinical Neurology (Vinken, P., Bruyn, G., eds.), Vol. 4, pp. 291—371. Amsterdam: North-Holland.

Hecaen, H. (1972): Introduction à la neuropsychologie. Paris: Larousse.

Hecaen, H., de Ajuriaguerra, J. (1963): Les Gauchers. Paris: P.U.F.

Hecaen, H., de Ajuriaguerra, J., David, M. (1952): Les déficits fonctionnels après lobectomie occipitale. Mschr. Psychiat. Neurol. *23*, 239—291.

Hecaen, H., Albert, M. L. (1978): Human Neuropsychology. New York: Wiley.

Hecaen, H., Angelergues, R. (1965): Neuropsychologie des dysfonctionnements des lobes occipitaux. Proc. 8th Int. Congr. Neurol. Vienna *3*, 29—45.

Hecaen, H., Angelergues, R., Dov Zenis, J. A. (1963): Les Agraphies. Neuropsychologia *1*, 179—208.

Hecaen, H., Consoli, S. (1973): Analysis of language troubles in the course of lesions of Broca's area. Neuropsychologia *11*, 377—388.

Hecaen, H., Dell, M. B., Roger, A. (1955): L'aphasie de conduction (Leitungsaphasie). L'Encéphale *2*, 170—195.

Hecaen, H., Dubois, J., Marcie, P. (1968): Les désorganisations de la réception des signes verbaux dans l'aphasie sensorielle. Revue d'Acoustique 287—304.

Hecaen, H., Dubois-Poulsen, M., Magis, G., Angelergues, R. (1952): Conséquences visuelles des lobectomies occipitales. Annales d'oculistique *185*, 305—347.

Hecaen, H., Marcie, P. (1974): Disorders of Written Language Following Right Hemisphere Lesions: Spatial Dysgraphia. In: Hemispheric Function in the Human Brain (Dimond, S., Beaumont, E., eds.), pp. 345—366. London: Paul Elek.

Hecaen, H., Maxurs, G., Ramier, A., et al. (1971): Aphasie croisée chez un sujet droitier bilingue. Rev. Neurol. *1*, 124, 319—323.

Hecaen, H., Penfield, W., Bertrand, C., Malmo, R. (1956): The syndrome of apractognosia due to lesions of the minor cerebral hemisphere. Arch. Neurol. Psychiat. *75*, 400—434.

Hecaen, H., Saugeuet, J. (1971): Cerebral dominance in left-handed subjects. Cortex *7*, 19—48.

Heilman, K. M., Coyle, J. M., Gonyea, E. E., Geschwind, N. (1973): Apraxia and agraphia in a left hander. Brain *96*, 21—28.

Heilman, K. M., Gonyea, E. F., Geschwind, N. (1974): Apraxia and agraphia in a right hander. Cortex *10*, 284—288.

Heilman, K. M., Roth, L., Campanella, D., Wolfson, S. (1979): Wernicke's and global aphasia without alexia. Arch. Neurol. *36*, 129—133.

Heilman, K. M., Scholes, R. J. (1976): The nature of comprehension errors in Broca's conduction and Wernicke's aphasics. Cortex *12*, 258—265.

Heilman, K. M., Scholes, R., Watson, R. T. (1975): Auditory affective agnosia. Disturbed comprehension of affective speech. J. Neurol. Neurosurg. Psychiat. *38*, 69—72.

Heilman, K. M., Scholes, R., Watson, R. T. (1976): Effects of immediate memory in Broca's and conduction aphasia. Brain Lang. *3*, 201—208.

Heilman, K. M., Tucker, D. M., Valenstein, E. (1976): A case of mixed transcortical aphasia with intact naming. Brain *99*, 415—426.

Helm, N. A. (1977): A program for stimulating recovery from agrammatism. Paper presented Amer. Sp. and Hear. Association Meeting, Chicago.

Helm, N. A. (1978): Criteria for selecting aphasia patients for melodic intonation therapy. Paper presented as part of symposium "Language Rehabilitation in Aphasia", American Academy for the Advancement of Science, Annual Meeting, Washington, D.C.

Helm, N. A., Benson, D. F. (1978): Visual Action Therapy for Global Aphasia. Paper presented to Academy of Aphasia Annual Meeting, Chicago.

Helm, N. A., Kaplan, E. F., Vercruysse, L. (1978): The role of gesture in naming. Unpublished study.

Helmick, J. W., Wipplinger, M. (1975): Effects of stimulus repetition on the naming behavior of an aphasic adult: a clinical report. J. Communic. Dis. *8*, 23—29.

Hemphill, R., Stengel, E. (1940): A study on pure word deafness. J. Neurol. Neurosurg. Psychiat. *3*, 251—262.

Henschen, S. E. (1918): Über die Hörsphäre. J. Psychol. Neurol. *22*, 30 (Ergänzungsheft 3), 319.

Henschen, S. E. (1922): Klinische und anatomische Beiträge zur Pathologie des Gehirns. Stockholm: Nordiske Bokhandeln.

Hier, D. B., Davis, K. R., Richardson, E. P., Mohr, J. P. (1977): Hypertensive putaminal hemorrhage. Annals Neurol. *1*, 152—159.

Hier, D. B., Mohr, J. P. (1977): Incongruous oral and written naming. Evidence for a subdivision of the syndrome of Wernicke's aphasia. Brain Lang. *4*, 115—126.

Hilpert, P. (1930): Die Bedeutung des linken Parietallappens für das Sprechen. J. Psychol. Neurol. *40*, 225—255.

Hoeft, H. (1957): Klinisch-anatomischer Beitrag zur Kenntnis der Nachsprechaphasie (Leitungsaphasie). Dtsch. Zschr. Nervenheilk. *175*, 560—594.

Holland, A. L. (1969): Some current trends in aphasic rehabilitation. ASHA *11*, 3—7.

Holland, A. L., Levy, C. B. (1971): Syntactic generalization in aphasics as a function of relearning an active sentence. Acta Symbolica 2, 34—41.

Howes, D. (1967): Some Experimental Investigations of Language in Aphasia. In: Research in Language Behavior and Some Neurophysiological Implications (Salzinger, K., Salzinger, S., eds.). New York: Academic Press.

Howes, D., Geschwind, N. (1964): Quantitative Studies of Aphasic Language. In: Disorders of Communication (Rioch, D. M., Weinstein, E. A., eds.), pp. 229—244. Baltimore: Williams and Wilkins.

Hubel, D. H., Wiesel, T. N. (1979): How the brain processes sensory information is suggested by studies of the primary visual cortex. Sci. Amer. *241*, 150—163.

Huber, M. (1944): A phonetic approach to the problem of perception in a case of Wernicke's aphasia. JSD *9*, 227—257.

Hyland, H. H. (1933): Thrombosis of intracranial arteries: Report of three cases involving respectively, the anterior cerebral, basilar, and internal carotid arteries. Arch. Neurol. Psychiat. *30*, 342—356.

Irigaray, L. (1967): Approche psycholinguistique du langage des déments. Neuropsychologia *5*, 25—52.

Irigaray, L. (1973): Le langage des déments. The Hague: Mouton.

Jackson, H. (1878): On the affections of speech from disease of the brain. Brain *1*, 304—330.

Jastak, J., Jastak, S. R. (1976): The Wide Range Achievement Test. (Revised.) Wilmington: Guidance Associates of Delaware.

Jerger, J., Lovering, L., Wertz, M. (1972): Auditory disorder following bilateral temporal lobe insult: report of a case. J. Sp. Hear. Dis. *37*, 523—535.

Jerger, J., Weikers, N., Sharbrough, F., Jerger, S. (1969): Bilateral lesions of the temporal lobe. Acta Oto-laryngol., Suppl. 258.

Johansson, T., Fahlgren, H. (1979): Alexia without agraphia: lateral and medial infarction of the occipital lobe. Neurol. *29*, 390—393.

Johns, D. F., Darley, F. I. (1970): Phonemic variability in apraxia of speech. J. Sp. Hear. Res. *13*, 556—583.

Johns, D. F., LaPointe, L. L. (1976): Neurogenic Disorders of Output Processing: Apraxia. In: Studies in Neurolinguistics (Whitaker, H., Whitaker, H. A., eds.), Vol. 1, Chapt. 5. New York: Academic Press.

Johnson, M. (1975): Recovery from aphasia following traumatic and surgical cortical lesions. Paper presented Amer. Sp. Hear. Assoc. Meeting, Washington, D.C.

Johnson, M. G., Rubens, A. B. (1975): Case Report: Visual-linguistic disturbances following left occipital lobectomy. Amer. Sp. Hear. Assoc. Meeting, Washington, D.C.

Johnson, M., Sahoske, P., Grembowski, C., Rubens, H. (1976): Preservation of responses requiring whole body movements in severe aphasia. Paper presented Amer. Sp. Hear. Assoc. Meeting, Houston, Tex.

Kaplan, E., Goodglass, H., Weintraub, S. (1978): The Boston Naming Test. Boston, Mass. (Unpublished experimental test.)

Kean, M. L. (1977): The linguistic interpretation of aphasic syndromes. Cognition *5*, 9—46.

Kerschensteiner, M., Jartje, W., Orgass, B., Poeck, K. (1972): The recognition of simple and complex realistic figures in patients with unilateral brain lesion. Arch. Psychiat. Nervenkhr. *216*, 188—200.

Kertesz, A., Harlock, W., Coates, R. (1979): Computer tomographic localization, lesion size and prognosis in aphasia and nonverbal impairment. Brain Lang. *8*, 34—50.

Kertesz, A., Lesk, D., McCabe, P. (1977): Isotope localization of infarcts in aphasia. Arch. Neurol. *34*, 590—601.

Kertesz, A., McCabe, P. (1977): Recovery patterns and prognosis in aphasia. Brain *100*, 1—18.

Kinsbourne, M. (1972): Behavioral analysis of the repetition deficit in conduction aphasia. Neurology 22, 1126—1132.

Kinsbourne, M., Rosenfeld, D. B. (1974): Agraphia selective for written spelling. Brain Lang. *1*, 215—225.

Kinsbourne, M., Warrington, E. K. (1964): Observations on colour agnosia. J. Neurol. Neurosurg. Psychiat. *27*, 296—299.

Klein, R., Harper, J. (1956): The problem of agnosia in the light of a case of pure word deafness. J. Ment. Sci. *102*, 112—120.

Kleist, K. (1934): Konstruktive (optische) Apraxie. In: Handbuch der ärztlichen Erfahrungen im Weltkriege (Bonhoeffer, K., ed.). Leipzig: Barth.

Kleist, K. (1934): Gehirnpathologie. Leipzig: Barth.

Kleist, K. (1962): Sensory aphasia and amusia. (Translated by Fish, T. J., Stanton, J. B.) Oxford: Pergamon Press.

Konorski, J., Kozniewski, H., Stepien, L. (1961): Analysis of symptoms and cerebral localization of the audio-verbal aphasia. Proceedings of the VIIth International Congress of Neurology II, pp. 234—236. Rome: Società Grafica Romana.

Krapf, E. (1957): A propos des aphasies chez les polyglottes. L'Encéphale 47, 623—629.

Kreindler, A., Ionasescu, V. (1961): A case of "pure" word blindness. J. Neurol. Neurosurg. Psychiat. *24*, 275—280.

Kurtzke, J. (1969): Epidemiology of Cerebrovascular Disease. Berlin-Heidelberg-New York: Springer.

Kussmaul, A. (1877): Disturbances of speech. Cyclop. Pract. Med. *14*, 581—875.

LaPointe, L. (1977): Base-10 programmed stimulation: task specification, scoring, and plotting performance in aphasia therapy. JSHD *42*, 90—105.

Larsen, B., Skinhoj, E., Lassen, N. A. (1978): Variations in regional cortical blood flow in the right and left hemispheres during automatic speech. Brain *101*, 193—210.

LeCours, A. R., Lhermitte, F. (1976): The "pure form" of the phonetic disintegration syndrome (pure anarthria); anatomical-clinical report of a historical case. Brain Lang. *3*, 88—113.

LeCours, A. R., Rovillon, F. (1976): Neurolinguistic analysis of jargon aphasia and jargon agraphia. In: Studies in Neurolinguistics (Avakian-Whitaker, H., Whitaker, H., eds.), Vol. 2, pp. 95—144. New York: Academic Press.

Leischner, A. (1943): Die „Aphasie" der Taubstummen. Archiv Psych. *115*, 469—548.

LeMay, M. (1976): Morphological cerebral asymmetries of modern man, fossil man and non-human primate. Ann. New York Academy of Science *280*, 349—366.

LeMay, M., Culebras, A. (1972): Human brain—morphologic differences in the hemispheres demonstrable by carotid arteriography. N. Engl. J. Med. *287*, 168—170.

LeMay, M., Geschwind, N. (1975): Hemispheric differences in the brains of great apes. Brain Behav. Evol. *11*, 48—52.

Lenneberg, E. H. (1967): Biological Foundations of Language. New York: J. Wiley.

Levine, D. N., Calvanio, R. (1978): A study of the visual defect in verbal alexia-simultan-agnosia. Brain *101*, 65—81.

Lezak, M. D. (1976): Neuropsychological Assessment. New York: Oxford University Press.

Lhermitte, F., Gautier, J. C. (1969): Aphasia. In: Handbook of Clinical Neurology (Vinken, R. R., Bruyn, G. W., eds.), pp. 84—104. Amsterdam: North-Holland.

Lhermitte, R., Hecaen, H., Dubois, J., *et al.* (1966): Le problème de l'aphasie des polyglottes. Neuropsychologia *4*, 315—329.

Lichtheim, L. (1884): Über Aphasie. Dtsch. Arch. klin. Med. *36*, 204—268.

Lichtheim, L. (1885): On aphasia. Brain *7*, 433—484.

Liepmann, H. (1898): Ein Fall von reiner Sprachtaubheit. In: Psychiatrische Abhandlungen (Wernicke, C., ed.). Breslau: Schletter Verlag.

Liepmann, H. (1900): Das Krankheitsbild der Apraxie („motorischen Asymbolie"). Mschr. Psychiat. Neurol. *8*, 15—44, 102—132, 181—197.

Liepmann, H. (1905): Die linke Hemisphäre und das Handeln. Münch. Med. Wschr. *2*, 2375—2378.

Liepmann, H., Maas, O. (1907): Fall von linksseitiger Agraphie und Apraxie bei rechtsseitiger Lähmung. J. Psychol. Neurol. *10*, 214—227.

Liepmann, H., Pappenheim, M. (1914): Über einen Fall von sogenannter Leitungsaphasie mit anatomischen Befund. Zschr. ges. Neurol. Psychiat. *27*, 1—41.

Ling, W., Gay, A. J. (1968): Optokinetic Nystagmus: A Proposed Pathway and Its Clinical Application. In: Neuro-Ophthalmology (Smith, J. L., ed.), Vol. 4, pp. 117—123. St. Louis: Mosby.

Luria, A. R. (1947): Traumatic Aphasia. Academy of Medical Sciences Press. Moscow.

Luria, A. (1963): Restoration of Function after Brain Injury. New York: Macmillan.

Luria, A. (1966): Human Brain and Psychological Processes. New York: Harper.

Luria, A. (1970): Traumatic Aphasia. The Hague: Mouton.

Luria, A. R., Hutton, J. T. (1977): A modern assessment of the basic forms of aphasia. Brain Lang. *4*, 129—151.

Luria, A. R., Tsvetkova, L. S. (1968): The mechanisms of "Dynamic Aphasia". (Foundation of Language, Vol. 4.) Amsterdam.

Magnan, D. R. (1880): On simple aphasia, and aphasia with incoherence. Brain *2*, 112—123.

Mahoudeau, D., Lemoyne, J., Foncin, J. R., Dubrisay, J. (1958): Considération sur l'agnosie auditive à propos d'un cas anatomoclinique. Rev. Neurol. *99*, 454—471.

Marie, P. (1906): Révision de la question de l'aphasie. Sem. Med. *21*, 241—247, 493—500, 565—571.

Marie, P., Foix, C. (1917): Les aphasies de guerre. Rev. Neurol. *24*, 53—87.

Marinesco, G., Grigoresco, D., Axentes, S. (1932): Aphasie croisée. Rev. Belge des Sci. Méd. *4*, 2—10.

Maroun, F. B., Jacob, J. C., Gowing, P. (1970): Dysphonia associated with cortical neoplasms. J. Neurosurg. *32*, 671—676.

Marshall, R. C., King, P. S. (1973): Effects of fatigue produced by isokinetic exercises on the communication ability of aphasic adults. JSHR *6*, 222—230.

Marshall, R. C., Newcombe, F. (1973): Patterns of paralexia: a psycholinguistic approach. J. Psycholinguistic Res. 2, 175—199.

Marshall, R. C., Watts, M. T. (1976): Relaxation training: effects on the communicative ability of aphasic adults. Arch. of Phys. Med. and Rehab. 57, 464—467.

Martin, A. D. (1974): Some objections to the term apraxia of speech. J. Sp. Hear. Dis. 39, 53—64.

Martin, A. D. (1977): Processing strategies in aphasia rehabilitation. Paper presented Amer. Sp. Hear. Assoc. Meeting, Chicago.

Masdev, J. C., Schoene, W. C., Funkenstein, H. (1978): Aphasia following infarction of the left supplementary motor area: a clinicopathologic study. Neurol. 28, 1220—1223.

Massopoust, L. C., Wolin, L. R., jr. (1967): Changes in auditory frequency discrimination thresholds after temporal cortex ablation. Exp. Neurol. 19, 245—251.

Mateer, C. (1978): Asymmetric effects of thalamic stimulation on rate of speech. Neuro-psychologia 16, 497—499.

Mateer, C., Kimura, D. (1977): Impairment of nonverbal oral movements in aphasia. Brain Lang. 4, 262—276.

Mattis, S., French, J. R., Rapin, I. (1975): Dyslexia in children and young adults: three independent neuropsychological syndromes. Develop. Med. Child Neurol. 17, 150—163.

McDearmon, J. R., Potter, R. E. (1975): The use of representational prompts in aphasia therapy. J. Commun. Dis. 8, 199—206.

Menyak, P. (1971): The Acquisition and Development of Language. Englewood Cliffs, N.J.: Prentice-Hall.

Miller, A. (1963): Verbal satiation and the role of concurrent activity. J. Abnor. Soc. Psych. 66, 206—212.

Mills, C. K. (1880): Aphasia. Philadelphia Med. Bull.

Mills, C. K. (1904): Treatment of aphasia by training. JAMA 43, 1940—1949.

Milner, B. (1964): Some Effects of Frontal Lobectomy in Man. In: The Frontal Granular Cortex and Behavior (Warren, J. M., Akert, K., eds.). New York: McGraw-Hill.

Minkowski, M. (1963): On Aphasia in Polyglots. In: Problems of Dynamic Neurology (Halpern, L., ed.). Jerusalem: Hebrew University Press.

Minkowski, M. (1965): Considérations sur l'aphasie des polyglottes. Rev. Neurol. 112, 486—495.

Mohr, J. P. (1976): Broca's Area and Broca's Aphasia. In: Studies in Neurolinguistics (Whitaker, H., Whitaker, H., eds.), Vol. 1. New York: Academic Press.

Mohr, J. P., Funkenstein, H., Finkelstein, S., Pessin, M., Duncan, G. W., Davis, K. (1975): Broca's area infarction versus Broca's aphasia. Neurology 25, 349.

Mohr, J. P., Caplan, L. R., Melski, J. R., et al. (1978): The Harvard Cooperative Stroke Registry: a prospective registry. Neurol. 28, 754—762.

Mohr, J. P., Hier, D. B., Kirshner, H. S. (1978): Modality bias in Wernicke's aphasia. Neurol. 4, 395. (Abstract.)

Mohr, J. P., Kase, C. S., Meckler, R. J., Fisher, C. M. (1977): Sensorimotor stroke due to thalamocapsular ischemia. Arch. Neurol. 34, 739—741.

Mohr, J. P., Pessin, M. S., Finkelstein, S., et al. (1978): Broca's aphasia: Pathologic and clinical aspects. Neurol. 28, 311—324.

Mohr, J. R., Sidman, M., Stoddard, L. T., et al. (1973): Evolution of the deficit in total aphasia. Neurol. (Minneap.) 23, 1302—1312.

Mohr, J. P., Watters, W. L., Duncan, G. W. (1975): Thalamic hemorrhage and aphasia. Brain Lang. 2, 3—17.

Moyer, S. B. (1979): Rehabilitation of alexia: a case study. Cortex 15, 139—144.

Muma, J. (1977): Language Intervention Strategies. Lang. Sp. Hear. Ser. in the Schools, Vol. 8, pp. 107—124.

Myers, R. E. (1976): Comparative Neurology of Vocalization and Speech: Proof of a Dichotomy. In: Origins and Evolution of Language and Speech (Harnad, S. R., Steklis, H., Lancaster, J., eds.). New York: The New York Academy of Sciences.

Naeser, M. A. (1975): A structured approach to teaching aphasics basic sentence types. Brit. J. Dis. Commun. 10, 70—76.

Naeser, M. A., Hayward, R. W. (1978): Lesion localization in aphasia with cranial computed tomography and the Boston Diagnostic Aphasia Exam. Neurol. *28*, 545—551.

Naeser, M. A., Levine, H. L., Helm, N., Laughlin, S. (1979): Putaminal aphasia—a unique aphasia syndrome. Presented at Academy of Aphasia.

Nebes, R. D. (1975): The nature of internal speech in a patient with aphemia. Brain Lang. *2*, 489—497.

Neisser, U. (1967): Cognitive Psychology. New York: Appleton.

Newcombe, F., Oldfield, R. C., Ratcliff, G. G., Wingfield, A. (1971): The recognition and naming of object-drawings by men with focal brain wounds. J. Neurol. Neurosurg. Psychiat. *34*, 329—340.

Newcombe, F., Oldfield, C., Wingfield, R. (1964): Object naming by dysphasic patients. Nature *207*, 1217—1218.

Nielsen, J. M. (1938): The unsolved problems in aphasia. Part I: Alexia in motor aphasia. Bull. LA Neurol. Soc. *4*, 114—122.

Nielsen, J. M. (1947): Agnosia, Apraxia, Aphasia, their value in cerebral localization. New York: Hoeber.

Nielsen, J. M., Raney, R. B. (1938): Symptoms following surgical removal of major (left) angular gyrus. Bull LA Neurol. Soc. *3*, 42—46.

Nielsen, J. M., Schutz, D. H., Corbin, M. L., Crittsinger, B. (1948): The treatment of traumatic aphasia of WWII at Birmingham Gen. V.A.H., Van Nuys, Calif. Milit. Surg. *10*, 351—364.

Niessl v. Mayendorf, E. (1911): Die aphasischen Symptome und ihre kortikale Lokalisation. Leipzig: Barth.

Obler, L. K., Albert, M. L., Goodglass, H., Benson, D. F. (1978): Aphasia type and aging. Brain Lang. *6*, 318—322.

Oldfield, R. C. (1966): Things, words and the brain. Quart. J. Exper. Psych. *18*, 340—353.

Olsen, W. O., Tillman, T. W. (1968): Hearing aids and sensorineural hearing loss. Ann. Otol. Rhin. Laryng. *77*, 717—726.

Ojemann, G. A. (1975): Language and the thalamus: Object naming and recall during and after thalamic stimulation. Brain Lang. *2*, 101—120.

Ojemann, G. A., Fedio, P. (1968): Effect of stimulation of the human thalamus and parietal and temporal white matter on short term memory. J. Neurosurg. *29*, 51—59.

Ojemann, G. A., Ward, A. A. (1971 b): Speech representation in ventrolateral thalamus. Brain *94*, 669—680.

Ojemann, G. A., Blick, K., Ward, A. A. (1971 a): Improvement and disturbance of short term verbal memory with human ventrolateral thalamus stimulation. Brain *94*, 225—240.

Ojemann, G. A., Fedio, P., Van Buren, J. M. (1968): Anomia from pulvinar and subcortical parietal stimulation. Brain *91*, 99—116.

Ojemann, G. A., Whitaker, H. A. (1978): Language localization and variability. Brain Lang. *7*, 239—260.

Oppenheimer, D. R., Newcombe, F. (1978): Clinical and anatomic findings in a case of auditory agnosia. Arch. Neurol. *25*, 712—719.

Osterrieth, P. A. (1944): Le test de copie d'une figure complexe. Arch. Psychol. *30*, 206—356.

Oxbury, J., Oxbury, S., Humphrey, N. (1969): Varieties of colour anomia. Brain *92*, 847—860.

Paradis, M. (1977): Bilingualism and Aphasia. In: Studies in Neurolinguistics (Whitaker, H., Whitaker, H. A., eds.), Vol. 3, Chapt. 2. New York: Academic Press.

Patterson, K. E., Marcel, A. J. (1977): Aphasia, dyslexia and the phonological coding of written words. Quart. J. Exper. Psychol. *29*, 307—318.

Pearlman, A. L., Birch, J., Meadows, J. C. (1979): Cerebral color blindness: An acquired defect in hue discrimination. Annals Neurol. *5*, 253—261.

Penfield, W., Roberts, I. (1959): Speech and Brain Mechanisms. Princeton, N.J.: Princeton University Press.

Penfield, W., Welch, K. (1951): The supplementary motor area of the cerebral cortex: a clinical and experimental study. A.M.A. Arch. Neurol. Psychiat. *66*, 289—317.

Pershing, H. T. (1900): A case of Wernicke's conduction aphasia with autopsy. J. Nerv. Ment. Dis. *27*, 369—374.

Petit-Dutaillis, D., Guiot, G., Messing, R., et al. (1954): À propos d'une aphémie par atteinte de la zone matrice supplémentaire de Penfield. Rev. Neurol. 90, 95—106.

Pick, A. (1898): Beiträge zur Pathologie und pathologischen Anatomie des Zentralnerven-systems, pp. 134—149. Berlin: Karger.

Pitres, A. (1895): Étude sur l'aphasie chez les polyglottes. Rev. Méd. 15, 873—899.

Pitres, A. (1898): L'aphasie amnésique et ses variétés cliniques, p. 94. Paris: Felix Alcan.

Poeck, K., Kerschensteiner, M. (1975): Analysis of the Sequential Motor Events in Oral Apraxia. In: Cerebral Localization (Zülch, K. J., Creutzgeldt, O., Galbraith, B. C., eds.). Berlin-Heidelberg-New York: Springer.

Poeck, K., Kerschensteiner, M., Hartje, W. (1972): A quantitative study on language under-standing in fluent and nonfluent aphasia. Cortex 8, 299—304.

Poppen, V. L. (1939): Ligation of the left anterior cerebral artery. Arch. Neurol. Psychiat. 41, 495—503.

Porch, B. (1971): Porch Index of Communicative Ability. Palo Alto: Consulting Psychologists Press.

Pötzl, O. (1925): Über die parietal bedingte Aphasie und ihren Einfluß auf das Sprechen mehrerer Sprachen. Zschr. ges. Neurol. Psychiatr. 96, 100—124.

Pötzl, O., Stengel, E. (1937): Über das Syndrom Leitungsaphasie-Schmerzasymbolie. Jahrb. Psychiat. 53, 174—207.

Prins, R. S., Snow, C. E., Wagenaar, E. (1978): Recovery from aphasia: Spontaneous speech versus language comprehension. Brain Lang. 6, 192—211.

Ramier, A. M., Hecaen, H. (1970): Rôle respectif des atteintes frontales et de la latéralisation lésionnelle dans les déficits de la "fluence verbale". Rev. Neurol. (Paris) 123, 17—22.

Reivich, M., Kuhl, D., Wolf, A., Greenberg, J., Phelps, M., I do I, Casella, V., Fowler, J., Hoffman, E., Alavi, A., Som, P., Sokoloff, L. (1979): The (18F) fluorodeoxyglucose method for the measurement of local cerebral glucose utilization in man. Circ. Res. 44, 127—137.

Reynolds, A. F., Harris, A. B., Ojemann, G. A., Turner, P. T. (1978): Aphasia and left thalamic hemorrhage. J. Neurosurg. 48, 570—574.

Reynolds, A. F., Turner, P. T., Harris, A. B., Ojemann, G. A., Davis, L. E. (1979): Left thalamic hemorrhage with dysphasia: A report of 5 cases. Brain Lang. 7, 62—73.

Ribot, T. (1881): Les maladies de la mémoire. Paris: Baillière.

Riklan, M., Levita, E. (1969): Subcortical Correlates of Human Behavior: A Psychological Study of Basal Ganglia and Thalamic Surgery. Baltimore: Williams and Wilkins.

Risse, G. L. (1978): The performance of aphasic patients on developmental conceptual tasks of Piaget. Paper presented at annual meeting International Neuropsychological Society, Minneapolis.

Robinson, B. W. (1976): Limbic Influences on Human Speech. In: Origins and Evolution of Language and Speech (Harnad, S. R., Senlis, H., Lancaster, J., eds.). New York: The New York Academy of Sciences.

Rosait, G., De Bastiani, P. (1979): Pure agraphia: a discrete form of aphasia. J. Neurol. Neurosurg. Psychiat. 42, 266—269.

Rosenbek, J. C., Green, E. F., Flynn, M., Wertz, R. T., Collins, M. (1977): Anomia: a clinical experiment. Paper presented to clinical aphasiology conference, Amelia Island, Florida.

Rosenbek, J. C., Lemme, M. L., Ahern, M. B., Harris, E. H., Wertz, R. T. (1973): A treatment for apraxia of speech in adults. JSHD 38, 462—472.

Ross, E. D., Mesulam, M. M. (1979): Dominant language functions of the right hemisphere. Arch. Neurol. 36, 144—148.

Rubens, A. B. (1975): Aphasia with infarction in the territory of the anterior cerebral artery. Cortex 11, 239—250.

Rubens, A. B. (1976): Transcortical Motor Aphasia. In: Studies in Neurolinguistics (Whitaker, H., Whitaker, H. A., eds.), Vol. 1, pp. 293—306. New York: Academic Press.

Russell, W. R., Espir, M. L. E. (1961): Traumatic Aphasia: A Study of Aphasia in War Wounds of the Brain. London: Oxford University Press.

Sabouraud, O., Gagnepain, J., Sabourand, A. (1963): Vers une approche linguistique des problèmes de l'aphasie (II), L'aphasie de Broca. Rev. Neuropsychiat. de L'Ouest 2, 3—38.

Saffran, E. M., Schwartz, M. F., Marin, O. (1976): Semantic mechanism in paralexia. Brain Lang. *3*, 255—265.

Samarel, A., Wright, T. L., Sergay, S., *et al.* (1976): Thalamic hemorrhage with speech disorder. Trans. Amer. Neurol. Assoc. *101*, 283—285.

Samuels, J. A., Benson, D. F. (1979): Some aspects of language comprehension in anterior aphasia. Brain Lang. *8*, 275—286.

Sands, E. S., Sarno, M. T., Shankweiler, D. P. (1969): Long-term assessment of language function in aphasia due to stroke. Arch. Phys. Med. Rehab. *50*, 202—207.

Sanides, F. (1970): Functional Architecture of Motor and Sensory Cortices in Primates in the Light of a New Concept of Neocortex Evolution. In: The Primate Brain (Noback, C. R., Montagna, W., eds.), pp. 137—208. New York: Appleton.

Sarno, M. T. (1974): Aphasia Rehabilitation. In: Communication Disorders, Remedial Principles and Practices (Dickson, S., ed.), pp. 339—440. Scott Foresman and Co.

Sarno, M. T. (1979): Outcome of language rehabilitation in the elderly aphasic patient. Paper presented to Conference on Language Communication in the Elderly, Boston.

Sarno, M. T., Levita, E. (1971): Natural course of recovery in severe aphasia. Arch. Phys. Med. *52*, 175—179.

Sarno, M. T., Silverman, M., Levita, E. (1970): Psychosocial factors and recovery in geriatric patients with severe aphasia. J. Amer. Geriat. Soc. *18*, 405—509.

Sarno, M. T., Silverman, M. G., Sands, E. S. (1970): Speech therapy and language recovery in severe aphasia. J. Sp. Hear. Res. *13*, 607—623.

Sarno, J. E., Swisher, L. P., Sarno, M. T. (1969): Aphasia in a congenitally deaf man. Cortex *5*, 398—414.

Sasanuma, S. (1975): Kana and Kanji processing in Japanese aphasics. Brain Lang. *2*, 369—383.

Schlanger, P. H., Schlanger, B. B. (1970): Adapting role-playing activities with aphasic patients. JSHD *35*, 229—235.

Schuell, H. M. (1953): Aphasic difficulties understanding spoken language. Neurol. *3*, 176—184.

Schuell, H. M., Carroll, V. S., Street, B. S. (1955): Clinical treatment of aphasia. JSHD *20*, 43—53.

Schuell, H. M., Jenkins, J. J., Jiménez-Pabón, E. (1964): Aphasia in Adults: Diagnosis, Prognosis, and Treatment. New York: Hoeber.

Schuster, P., Taterka, H. (1926): Beitrag zur Anatomie und Klinik der reinen Worttaubheit. Ztschr. Neurol. Psychiat. *105*, 494—538.

Schwab, O. (1927): Über Stützreaktionen (Magnus) beim Menschen. (Zugleich ein Beitrag zur Auffassung der sogenannten Gelenksreflexe.) Ztschr. ges. Neurol. Psychiat. *108*, 585—593.

Schwartz, M. P., Saffran, E. M., Marin, O. S. M. (1978): The nature of the comprehension deficit in agrammatic aphasics. Paper presented at the Sixth Annual Meeting of the International Neuropsychology Society in Minneapolis, Minn., February 1978.

Scoresby-Jackson, R. (1867): Case of aphasia with right hemiplegia. Edinburgh Med. J. *12*, 696—706.

Seglas, J. (1892): Les troubles du langages chez les aliénés. Paris: Rueff.

Selby, G. (1967): Stereotaxic surgery for the relief of Parkinson's disease. II. An analysis of the results of a series of 303 patients (413 operations). J. Neurol. Sci. *5*, 343—375.

Shallice, T., Warrington, E. K. (1977): The possible role of selective attention in acquired dyslexia. Neuropsychologia *15*, 31—41.

Silverberg, R., Albert, M. L. (1975): Intellectual functioning in two geriatric groups: normal and Parkinsonian (abstract). Proc. 10th Internat. Congr. Gerontol. *2*, 109.

Silverberg, R., Gordon, H. (1978): Differential aphasia in two bilinguals. Neurol. *29*, 51—56.

Singer, H. D., Low, A. A. (1933): The brain in a case of motor aphasia in which improvement occurred with training. Arch. Neurol. Psychiat. *29*, 162—165.

Sjogren, T., Sjogren, H., Lindgren, A. (1952): Morbus Alzheimer and Morbus Pick. A genetic clinical and pathoanatomical study. Acta Psychiat. Neurol. Scand., Suppl. 82.

Skelly, M., Schinsky, L., Smith, R. W., Fust, R. S. (1974): American indian sign (Amerind) as a facilitator of verbalization for the oral verbal apraxic. JSHD *39*, 445—456.

Smith, A. (1971): Objective indices of severity of chronic aphasia in stroke patients. JSHD *36*, 167—207.

Smyth, G. E., Stern, K. (1938): Tumors of the thalamus. Brain *61*, 339—360.

Sparks, R., Helm, N., Albert, M. (1974): Aphasia rehabilitation resulting from melodic intonation therapy. Cortex *10*, 303—316.

Sparks, R., Holland, A. (1976): Method: Melodic intonation therapy for aphasia. JSHD *41*, 287—297.

Spellacy, F. (1970): Lateral preferences in the identification of patterned stimuli. J. Acoust. Soc. Amer. *47*, 574—578.

Spreen, O., Benton, A., Fincham, R. (1965): Auditory agnosia without aphasia. Arch. Neurol. *13*, 84—92.

Spreen, O., Benton, A., Van Allen, M. (1966): Dissociation of visual and tactile naming in amnestic aphasia. Neurology *16*, 807—814.

Sroka, H., Solsi, P., Bornstein, B. (1973): Alexia without agraphia with complete recovery. Confinia Neurol. *35*, 167—176.

Stachowiak, F. J., Poeck, K. (1976): Functional disconnection in pure alexia and color naming deficit demonstrated by facilitation methods. Brain Lang. *3*, 135—143.

Staller, J., Buchanon, D., Singer, M., Lappin, J., Webb, W. (1978): Alexia without agraphia: an experimental case study. Brain Lang. *5*, 378—387.

Stengel, E. (1933): Zur Lehre von der Leitungsaphasie. Ztschr. ges. Neurol. Psychiat. *149*, 266—291.

Stengel, E. (1936): Zur Lehre von den transkorticalen Aphasien. Ztschr. ges. Neurol. Psychiat. *154*, 778—782.

Stengel, E. (1947): A clinical and psychological study of echo-reactions. J. Ment. Sci. *93*, 598—612.

Stengel, E. (1964): Psychopathology of dementia. Proc. Rog. Soc. Med. *57*, 911—914.

Stengel, E., Lodge Patch, I. C. (1955): "Central" aphasia associated with parietal symptoms. Brain *78*, 401—416.

Stoicheff, M. L. (1960): Motivating instructions and language performance of dysphasic subjects. JSHR *3*, 75—85.

Stone, L. (1934): Paradoxical symptoms in right temporal tumor. J. Nerv. Ment. Dis. *79*, 1—13.

Strub, R., Gardner, H. (1974): The repetition deficit in conduction aphasia: Amnestic or linguistic? Brain Lang. *1*, 241—257.

Subirana, A. (1969): Handedness and Cerebral Dominance. In: Handbook of Clinical Neurology (Vinken, P., Bruyn, G., eds.), Vol. 4. Amsterdam: North-Holland.

Sweet, W. (1951): Discussion of Erickson and Woolsey. Trans. Amer. Neurol. Assoc. *76*, 55.

Taylor, M. (1965): A measurement of functional communication in aphasia. Arch. Phys. Med. Rehab. *47*, 101—106.

Terry, R. (1976): Dementia: A brief and selective review. Arch. Neurol. *33*, 1—4.

Thurstone, I. L., Thurston, T. (1949): Examiner Manual for the SRA Primary Mental Abilities (rev. ed.). Chicago: Illinois Scientific Research Association.

Trost, J. R., Canter, G. J. (1974): Apraxia of speech in patients with Broca's aphasia: A study of phoneme production accuracy and error patterns. Brain Lang. *1*, 63—79.

Truex, R. C., Carpenter, M. B. (1969): Human Neuroanatomy. Baltimore: Williams and Wilkins.

Tureen, K., Smolik, E., Tritt, J. (1951): Aphasia in a deaf-mute. Neurol. *1*, 237—249.

Tzortis, C., Albert, M. L. (1974): Impairment of memory for sequences in conduction aphasia. Neuropsychol. *12*, 355—366.

Ulatowskia, H. R., Richardson, S. M. (1974): A longitudinal study of an adult with aphasia: considerations for research and therapy. Brain Lang. *1*, 151—166.

Ulrich, G. (1978): Interhemispheric functional relationships in auditory agnosia. An analysis of the preconditions and a conceptual model. Brain Lang. *5*, 286—300.

Valenstein, E., Heilman, K. M. (1979): Apraxic agraphia with neglect-induced paragraphia. Arch. Neurol. *36*, 506—509.

Van Buren, J. M. (1975): The question of thalamic participation in speech mechanisms. Brain Lang. *2*, 31—44.

Victor, M., Adams, R. D., Collins, G. H. (1971): The Wernicke-Korsakoff Syndrome. Philadelphia: F. A. Davis.

Vignolo, L. (1964): Evolution of aphasia and language rehabilitation. Cortex *1*, 344—352.

Vignolo, L. (1969): Auditory agnosia: A review and report of recent evidence. In: Contributions to Clinical Neuropsychology (Benton, A., ed.), pp. 172—208. Chicago: Aldine.

Vincent, F. M., Sadowsky, C. H., Saunders, R. L., Reeves, A. G. (1977): Alexia without agraphia, hemianopia or color-naming defect: A disconnection syndrome. Neurol. *27*, 689—691.

Von Stockert, T. R. (1974): Ein neues Konzept zum Verständnis der cerebralen Sprachstörungen. Nervenarzt *45*, 94—97.

Wada, J. A., Clarke, R., Hamm, A. (1975): Cerebral hemispheric asymmetry in humans. Arch. Neurol. *32*, 239—246.

Wagenaar, E., Snow, C., Prins, R. (1975): Spontaneous speech of aphasia patients: A psycholinguistic analysis. Brain Lang. *2*, 281—303.

Walshe, T. M., Davis, K. R., Fisher, C. M. (1977): Thalamic hemorrhage: a computer tomographic-clinical correlation. Neurol. *27*, 217—222.

Warrington, E. K., Logue, V., Pratt, R. T. C. (1971): The anatomical localization of selective impairment of auditory verbal short-term memory. Neuropsychologia *9*, 377—387.

Warrington, E. K., Shallice, T. (1969): The selective impairment of auditory verbal short-term memory. Brain *92*, 885—896.

Warrington, E. K., Shallice, T. (1972): Neuropsychological evidence of visual storage in short-term memory tasks. Quart. J. Exp. Psychol. *24*, 30—40.

Watson, R. T., Heilman, K. M. (1979): Thalamic neglect. Neurol. *29*, 690—694.

Weidner, W. E., Lasky, E. Z. (1976): The interaction of rate and complexity of stimulus on the performance of adult aphasic subjects. Brain Lang. *3*, 34—40.

Weigel-Crump, C., Koenigsknecht, R. A. (1973): Tapping the lexical store of the adult aphasic analysis of the improvement made in word retrieval skills. Cortex *9*, 411—418.

Weigl, E. (1968): On the Problem of Cortical Syndromes: Experimental Studies. Reprinted from: The Reach of Mind (Simmel, M. L., ed.). New York: Springer.

Weigl, E., Bierwisch, M. (1973): Neuropsychology and Linguistics: Topics of Common Research. In: Psycholinguistics and Aphasia (Goodglass, H., Blumstein, S., eds.), pp. 1—28. Baltimore: Johns Hopkins University Press.

Weinstein, S. (1959): Experimental analysis of an attempt to improve speech in cases of expressive aphasia. Neurol. *9*, 632—635.

Weisenburg, T. S., McBride, K. (1935): Aphasia: A clinical and psychological study. New York: The Commonwealth Fund.

Wepman, J. M. (1953): A conceptual model for processes involved in recovery from aphasia. JSHD *18*, 4—13.

Wepman, J. M., Bock, R. D., Jones, L., Van Pelt, D. (1956): Psycholinguistic study of aphasia: A revision of the concept of anomia. J. Sp. Hear. Dis. *21*, 468—477.

Wepman, J. M., Jones, L. B. (1961): The Language Modalities Test for Aphasia. Chicago: Education Industry Service.

Wernicke, C. (1874): Der aphasische Symptomenkomplex. Breslau: Cohn und Weigart.

Wernicke, C. (1885—86): Neuere Arbeiten über Aphasie. Fortschritte der Medizin (cited by Dejerine).

Wernicke, C. (1908): The Symptom-Complex of Aphasia. In: Modern Clinical Medical Diseases of the Nervous System (Church, A., ed.). New York: Appleton.

Wertz, R. (1978): The effects of treatment on recovery from aphasia. Paper presented as part of symposium "Language Rehabilitation in Aphasia", American Academy Advancement of Science Annual Meeting, Washington, D.C.

West, J. (1977): Heightening visual imagery in aphasia treatment. Paper presented Amer. Sp. and Hear. Assoc. Meeting, Chicago.

Whitaker, H. (1976): A Case of the Isolation of the Language Function. In: Studies in Neurolinguistics (Whitaker, H., Whitaker, H. A., eds.), Vol. 1. New York: Academic Press.

Whitaker, H. A., Noll, J. D. (1972): Some linguistic parameters of the token test. Neuropsychologia *10*, 395—404.

Wilson, S. A. K. (1908): A contribution to the study of apraxia. Brain *31*, 164—216.

Wilson, S. A. K. (1926): Aphasia. London: Trubner and Co.

Woerkom, W. van (1925): Über Störungen im Denken bei Aphasiepatienten. Mschr. Psychiat. Neurol. *59*, 256—322.

Wolfart, G., Lindgren, A., Jernelius, B. (1952): Clinical picture and morbid anatomy in a case of "pure word deafness". J. Nerv. Ment. Dis. *115*, 116—118.

Yamadori, A. (1975): Ideogram reading in alexia. Brain *98*, 213—238.

Zangwill, O. (1960): Cerebral dominance and its relation to psychological functioning. London: Oliver and Boyd.

Zeki, S. M. (1973): Colour coding in Rhesus monkey prestriate cortex. Brain Res. *53*, 422—427.

Zurif, E., Caramazza, A., Myerson, R. (1972): Grammatical judgements of agrammatic patients. Neuropsychologia *10*, 405—417.

Zurif, E., Green, E., Caramazza, A., Goodenough, L. C. (1976): Grammatical intuitions of aphasic patients: Sensitivity to functors. Cortex *12*, 183—186.

# Subject Index